YOUR KNOWLEDGE HAS VALUE

- We will publish your bachelor's and master's thesis, essays and papers

- Your own eBook and book - sold worldwide in all relevant shops

- Earn money with each sale

Upload your text at www.GRIN.com and publish for free

Bibliographic information published by the German National Library:

The German National Library lists this publication in the National Bibliography; detailed bibliographic data are available on the Internet at http://dnb.dnb.de .

Imprint:

Copyright © 2016 GRIN Verlag, Open Publishing GmbH
Print and binding: Books on Demand GmbH, Norderstedt Germany
ISBN: 9783668480841

This book at GRIN:

http://www.grin.com/en/e-book/370769/studies-on-genetic-relationships-among-locally-cultivated-citrus-varieties

Prem Jose Vazhacharickal, Sajeshkumar N. K., Jiby John Mathew, Jaina James

Studies on genetic relationships among locally cultivated Citrus varieties in Kerala employing matK and rbcL gene using PCR technique and RFLP markers

GRIN Publishing

Studies on genetic relationships among locally cultivated Citrus varieties in Kerala employing matK and rbcL gene using PCR technique and RFLP markers

Prem Jose Vazhacharickal, Sajeshkumar N.K, Jiby John Mathew and Jaina James

ACKNOWLEDGEMENTS

Firstly we thank **God Almighty** whose blessing were always with us and helped us to complete this project work successfully.

We wish to thank our beloved Manager **Rev. Fr. Dr. George Njarakunnel,** Respected Principal **Dr. Joseph V. J,** Vice Principal **Fr. Joseph Allencheril**, Bursar **Shaji Augustine** and the Management for providing all the necessary facilities in carrying out the study. We express our sincere thanks to **Mr. Binoy A Mulanthra** (lab in charge, Department of Biotechnology) for the support. This research work will not be possible with the co-operation of many farmers.

We are gratefully indebted to our teachers, parents, siblings and friends who were there always for helping us in this project.

Prem Jose Vazhacharickal*, Sajeshkumar N.K, Jiby John Mathew and Jaina James

Table of contents

Table of figures

List of abbreviations

°C	: Degree celeseus
AFLP	: Amplified fragment length polymorphism
BLAST	: Basic local alignment search tool
Bp	: Base pairs
Cm	: Centimeter
CO$_2$	: Carbon dioxide
cpDNA	: Chloroplast DNA
CpDNA	: Chloroplast DNA
CTAB	: Cetyl trimethyl-ammonium bromide
Da	: Dalton
dNTP's	: Deoxynucleotides
EDTA	: Ethylene diamine tetra acetic acid
EST	: Expressed sequence tags
GPS	: Global positioning system
HDL	: High density lipoproteins
KDa	: Kilo Dalton
L	: Liter
LDL	: Low density lipoprotein
M	: Molar
matK	: Maturase K
MEGA	: Molecular evolution genetic analysis
mg/mL	: Milligram per milli liter
mM	: Milli molar
NCBI	: National center for biotechnology information
NJ	: Neighbour joining
Nm	: Nano meter
OD	: Optical density
PCR	: Polymerase Chain Reaction
PVP	: Polyvinyl pyrrolidone
RAPD	: Random amplified polymorphic DNA
rbcL	: Ribulose-1,5- bisphosphate carboxylase oxygenase
RFLP	: Restriction fragment length polymorphism

RNA	: Ribonucleic acid
RuBisCO	: Ribulose-1, 5-bisphosphate carboxylase oxygenase
SNP	: Single nucleotide polymorphism
SRAP	: Sequence related amplified polymorphism
SSR	: Simple sequence repeat
TE	: Tris EDTA
WBC	: White blood cells
µg	: Micro gram
µL	: micro litre

Studies on genetic relationships among locally cultivated Citrus varieties in Kerala employing matK and rbcL gene using PCR technique and RFLP markers

Prem Jose Vazhacharickal[1]*, Sajeshkumar N.K[1], Jiby John Mathew[1] and Jaina Joseph[1]

* premjosev@gmail.com

[1]Department of Biotechnology, Mar Augusthinose College, Ramapuram, Kerala, India-686576

Abstract

Citrus, one of the major genes of Rutaceae family and most economically important fruit tree and widely cultivated throughout the country. The Citrus have high nutritional value and medicinal value. The three varieties obtained from various districts in Kerala were used in this study. The phylogenetics is the study of the evolutionary relationships among different species which have common ancestor. These relationships are shown in the form of phylogenetic trees composed of branches which indicate the descendents and nodes which represent the most recent common ancestors. In order to assess the phylogenetic relationships among the 6 samples belongs to the 3 Citrus species (*Citrus aurantifolia*, Citrus *maxima*, *Citrus limon*) a chloroplast gene rbcL and matK was successfully amplified and sequenced. For this study DNA was extracted by using CTAB method. This extracted DNA was analysed by spectrophotometry method for checking purity of DNA.The samples were gel electrophoresed by 1% agarose gel electrophoresis at 80 volts. After electrophoresis the gel is examined in gel documentation system. The DNA band was observed under UV light looking florescent Orangish red colour. The extracted DNA was amplified by PCR method and PCR sample was applied for electrophoresis for checking DNA bands. PCR sample were purified sent for sequencing. Sequences obtained were subjected to editing and alignment using ClustalW programme of Bio Edit and phylogenetic analysis was performed using MEGA software. High similarity observed between selected verities. The locally collected gene sequence-based phylogeny presented here provides support for the early studies of speciation within the Rutaceae. An understanding of the main phylogenetic relationships between Citrus species will help to fine-tune the taxonomy of Rutaceae.

Keywords: Anti-oxidant; matK; Citrus; PCR; RFLP; Underutilized fruit.

1. Introduction

The genus Citrus, which includes mandarin, orange, lemon, grapefruit and lime, has high economic and nutritional value. This genus belongs to the subfamily Aurantioideae, which is one of the 7 subfamilies of the family Rutaceae. Therefore, phylogenetic study of both the genus Citrus and of the subfamily Aurantioideae is important (Penjor et al., 2013). The Aurantioideae consists of 2 tribes with 33 genera (Swingle et al., 1967). These 2 tribes are each composed of 3 subtribes: the tribe Clauseneae, which includes Micromelinae, Clauseninae, and Merrillinae; and the tribe Citreae, which includes Triphasiinae, Citrinae, and Balsamocitrinae. None of the Clauseneae species develop axillary spines, and the odd-pinnate leaves have alternately attached leaflets. The fruits are usually small and carry semi-dry or juicy berries, except in Merrilli (Swingle et al., 1967). In contrast, nearly all the species develop axillary spines in the Citreae. The simple leaves are easily distinguished from those of the tribe Clauseneae (Swingle et al., 1967). The genus Citrus belongs to the "true citrus fruit trees." The characteristics of Citrus species include asexual reproduction, high mutation frequency, and cross compatibility between species. Because of these characteristics, there is great morphological and ecological diversity among Citrus species Citrus species can be classified into 3 clusters: a citron cluster, a pummelo cluster, and a mandarin cluster. The samples used in this study belong to pummelo cluster (Penjor et al., 2013).

Since the 1970s, morphological and biochemical studies have been conducted to elucidate the phylogeny of Aurantioideae, especially of Citrus and its close relatives. Because of improved DNA analysis, these relationships have been studied extensively. Several techniques, such as restriction fragment length polymorphism (RFLP), random amplified polymorphic DNA (RAPD), simple sequence repeat (SSR), and sequence related amplified polymorphisms (SRAP) have been commonly used in taxonomic studies (Barkley et al., 2006; Algabal et al., 2011). Citrus are widely described and cultivated worldwide. Citrus species are small to medium sized shrubs or trees that are cultivated throughout the tropics and sub tropics. Citrus is primarily valued for the fruit , which is either eaten alone as fresh fruit, processed into juice, or added to dishes and beverages. All species have traditional medicinal value. Due to their universal appearance and easiness in harvesting, medicinal value Citrus are considered as standard specimens for several scientific studies including tissue culture and molecular phylogenetics. All these factors have contributed for the selection of *Citrus aurantifolia, Citrus maxima, Citrus limon* as a principle samples for the isolation of plant genomic DNA and for the molecular phylogenetics in a comparative sense to aid this study. The phylogenetics is the study of the evolutionary relationships among different species which have common ancestor. These relationships are shown in the form of phylogenetic trees composed of branches which indicate the descendents and nodes which represent the most recent common ancestors.

Phylogenetic approaches have been used on the sequences of DNA and proteins to determine the ancestral relationships of living organisms in the form of tree of life (Philippe and Forterre, 1999). These sequence-based approaches have also been used to study the relationships among different viruses and considered more reliable compared to other non sequence-based approaches (Bawden et al., 2000; McGeoch et al., 1995; Davison, 2002). The gene used for this study is rbcL (Ribulose-1,5-bisphosphate carboxylase oxygenase) and matK (Maturase K) gene. RuBisco (Ribulose-1,5- bisphosphate carboxylase oxygenase), is an enzyme found in chloroplast of plant cells that functions in the Calvin cycle to fix carbon dioxide (CO_2) and help drive the synthesis of glucose. The matK gene is also located on the chloroplast DNA (cpDNA) and encodes a maturase involved in splicing type II introns from RNA transcripts. The matK gene is encoded by the chloroplast trnK intron (Hilu and Liang, 1997).

1.1 Objectives

The objective of this study was polymerase chain reaction (PCR) based amplification, and sequencing of *Citrus aurantifolia, Citrus maxima, Citrus limon* and determination of the phylogenetic relationships based on matK gene and rbcL gene. Because the development of DNA-based genetic markers has had a revolutionary impact on plant genetics. It is theoretically possible to observe and exploit genetic variation in the entire genome of organisms with DNA markers. Ribulose-1,5- bisphosphate carboxylase oxygenase, is an enzyme found in chloroplast of plant cells that functions in the Calvin cycle to fix CO_2 and help drive the synthesis of glucose. RuBisco is an important enzyme in photosynthesis because without it plants would not be able to form glucose and other necessary metabolites to sustain life. Research has shown that decreases in

RuBisco can dramatically alter plant size and thus leading to smaller crop yields. The other gene used in this study is matK gene. The matK gene is also located on the cpDNA and encodes a maturase involved in splicing type II introns from RNA transcripts. The matK gene is encoded by the chloroplast trnK intron. Since matK has a relatively fast mutation rate, it evolves faster than the rbcL gene therefore; matK analysis should be useful for studying the phylogeny of different genera. The matK gene, formerly known as orfK, is emerging as yet another gene with potential contributions to plant systematics and evolution (Hilu and Liang, 1997). For this experiment rbcL primer and matK primer is selected for amplifying the DNA of the plants stated above and their sequences are analyzed for similarities and dissimilarities to check the evolutionary relationships amongst them.

1.2 Scope of the study

1.   To generate a base line information on the phylogenetic relatedness between the locally available Citrus varieties viz. *Citrus aurantifolia*, *Citrus maxima* and *Citrus limon*.

2.   To understand the variations among the said varieties of Citrus and also with other selected members of the Citrus family.

Gene specific primers used here are matK and rbcL gene. The matK gene was first identified by Sugita et al. (1985) from tobacco (*Nicotiana tabacum*) when they sequenced the trnK gene encoding the tRNA (Lys) of the chloroplast and rbcL. The rbcL gene is a valuable tool for assessing phylogenetic relationships. This gene is found in the chloroplasts of most photosynthetic organisms. These genes are analysed by PCR amplification and sequence analysis. This is based on the principle that Denaturation at 94°C, Annealing at 54°C and extension at 72°C .Temperature and time may depend on the primer used (Chen et al., 2002).

1.3 Taxonomical classification

Kingdom: Plantae-- planta, plantes, plants, vegetal

Subkingdom: Viridiplantae

Division: Tracheophyta – vascular plants, tracheophytes

Class: Magnoliopsida

Order: Sapindales

Family: Rutaceae - rues

Genus: Citrus L. - citrus

Species: *Citrus aurantifolia/Citrus maxima/Citrus limon*

2. Review of literature

Citrus is a genus of family Rutaceae that comprise some 158 genera and 1900 species (Mabberley, 2008). It is mainly tropical to semi tropical in origin and is assumed to have originated from the region within Northeast India, South China, Indonesia and Peninsular Malaysia (Swingle et al., 1967; Timmer et al., 1988; Cunningham and Harden, 1998). Citrus grows particularly well in areas where there is enough rainfall or irrigation to maintain growth and freezing conditions are not severe enough to kill the tree (Timmer et al., 1988; Cunningham and Harden, 1998). Citrus is also one of the most important fruit crops in the world and its international production has reached 122 million tons (FAO, 2008).

Citrus fruits and juices are a great source of bioactive compounds including antioxidants such as ascorbic acid, flavonoids, phenolic compounds and pectins that are important to human nutrition (Ghasemi et al., 2009; Fernandez-Lopez et al., 2005; Jayaprakasha and Patil, 2007). The peel which represents almost one half of the fruit mass has the highest concentrations of flavonoids in the Citrus fruit (Manthley and Grohmann, 1996; Manthley and Grohmann, 2001; Anagnostopoulou et al., 2006). A wide range of DNA markers is available and has been used to study the classification of Citrus genus, and phylogenetic relationships within Citrus and with related genera. These molecular studies have provided some insight to Citrus phylogeny (Wali et al., 2013). The phylogeny and taxonomy of Citrus fruit are complex, confusing and controversial due to the genetic heterogeneity of the genes, as well as its polyembryonic nature and the long generation time needed to carry out selection and recombination (Swingle et al., 1996).

2.1 Uses and importance

Citrus aurantifolia/lime: -.The *Citrus aurantifolia* have antioxidant effect. The juice and peel extract of *Citrus aurantifolia* have antioxidant properties. Antioxidant effects of *Citrus aurantifolia* juice and peel extract on low-density lipoprotein (LDL) oxidation. The principal use is still for food, refreshing drinks, tasty desserts, and for seasoning meats, vegetables, salads, sauces, and casseroles (Ehler, 2002; Katzer, 2002). Essential oils of Key lime and some other Citrus fruits cause phytophotodermatosis in sensitive individuals (Bruneton, 1999). Key lime is used to treat a huge number of ailments (Burkill 1997). It is a good honey plant. The plant can be used for a living fence post (Little and Wadsworth1964) and can be formed into a hedge (Burkill, 1997).

Citrus maxima or *Citrus grandis*/pommelo: -.They used for hyperlipidemia, atherosclerosis, reducing hematocrit counts, cancer, psoriasis, and for weight loss and obesity. Seed extract is used orally for bacterial, viral, and fungal infections including yeast infections, *Giardia lamblia* and *Entamoeba histolytica* (Orwa et al., 2009). Oil is used for muscle fatigue, hair growth, toning the skin, and for acne and oily skin. It is also used for the common cold, swine flu, and flu (influenza). Seed extract is used topically as a facial cleanser, first-aid treatment, as a treatment for mild skin irritations, and as a vaginal douche for vaginal candidiasis (yeast infection). It is also used as an ear or nasal rinse for preventing and treating infections; as a gargle for sore throats; and a dental rinse for preventing gingivitis, promoting healthy gums, and as a breath freshener. In food and beverages, it is consumed as a fruit, juice, and is used as a flavouring component (Orwa et al., 2009).

Citrus limon/lemon: - It helps in production of white blood cells (WBC) and antibodies in blood which attacks the invading microorganism and prevents infection. Lemon is an antioxidant which deactivates the free radicals preventing many dangerous diseases like stroke, cardiovascular diseases and cancers. Lemon lowers blood pressure and an increase the levels of high-density lipoprotein (HDL). It is found to be anti-carcinogenic which lowers the rates of colon, prostate and breast cancer (Gulsen and Roose, 2001).They prevent faulty metabolism in the cell, which can predispose a cell to becoming carcinogenic. Lemon juice is used to prevent common cold. Lemon juice is given to prevent/treat urinary tract infection Gonorrhoea. Lemon juice relieves colic pain and gastric problems. Lemon juice soothes the dry skin when applied with little glycerine. Lemon juice used for marinating seafood or meat kills bacteria and other organisms present in them, thereby prevents many gastric-intestinal tract infections (Penniston et al., 2008).

2.2 Molecular markers

The development of DNA-based genetic markers has had a revolutionary impact on animal genetics. It is theoretically possible to observe and exploit genetic variation in the entire genome of organisms with DNA markers. Allozymes, mitochondrial DNA, RFLP, RAPD, amplified fragment length polymorphism (AFLP), microsatellite, single nucleotide polymorphism (SNP), and expressed sequence tag (EST) markers are the popular genetic markers employed in genomic studies (Aljanabi et al., 1997).

One of the main questions at the beginning of any genome study is what type of marker is most suitable for the given project and to the species of interest. There is no simple answer to this question, and much depends on the specific objectives of the study. Depending on the problem addressed, on the available resources, time and costing, some molecular markers can be more appropriate than others for studying a given problem. For example, DNA sequencing provides a resolution appropriate to phylogenetic and population-level studies but microsatellites are more appropriate to studies of parentage and breeding (Allendorf et al., 2010).

2.2.1 Chloroplast DNA analysis

In these study two genes, namely, rbcL and matK were used to elucidate the phylogenetic relatedness of the selected species. The gene rbcL encodes Ribulose-1, 5-bisphosphate carboxylase oxygenase, commonly known by the abbreviation RuBisCO, is an enzyme involved in the first major step of carbon fixation, a process by which atmospheric carbon dioxide is converted by plants to energy rich molecules such as glucose. The rbcL gene, located on the cpDNA (Penjor et al., 2013). Compared to most genes encoded in the cpDNA, the rbcL gene has a relatively slow nucleotide substitution rate. In plants, algae, cyanobacteria, and phototrophic and chemoautotrophic proteobacteria, the enzyme usually consists of two types of protein subunit, called the large chain (L, about 55,000 Da) and the small chain (S, about 13,000 Da).

The large-chain gene is part of the chloroplast DNA molecule in plants.]There are typically several related small-chain genes in the nucleus of plant cells, and the small chains are imported to the stromal compartment of chloroplasts from the cytosol by crossing the outer chloroplast membrane. The enzymatically active substrate (ribulose1, 5-bisphosphate) binding sites are located in the large chains that form dimmers in which amino acids from each large chain contribute to the binding sites (Penjor et al., 2013). A total of eight large-chains (= 4 dimmers) and eight small chains assemble into a larger complex of about 540,000Da .In some proteobacteria and dinoflagellates, enzymes consisting of only large subunits have been found. Magnesium ions (Mg^{2+}) are needed for enzymatic activity. Correct positioning of Mg^{2+} in the active site of the enzyme involves addition of an "activating" carbon dioxide molecule (CO_2) to a lysine in the active site (forming a carbamate). Formation of the carbamate is favoured by an alkaline pH. The pH and the concentration of magnesium ions in the fluid compartment (in plants, the stroma of the chloroplast) increases in the light.

The other gene used for this study is matK gene. The matK gene is also located on the cpDNA and encodes a maturase involved in splicing type II introns from RNA transcripts. The matK gene is encoded by the chloroplast trnK intron. Since matK has a relatively fast mutation rate, it evolves faster than the rbcL gene .Therefore; matK analysis should be useful for studying the phylogeny of the genera included Aurantioideae. The matK gene, formerly known as orfK, is emerging as yet another gene with potential contributions to plant systematics and evolution (Johnson and Soltis, 1994; Johnson and Soltis, 1995; Steele and Vilgalys, 1994; Liang and Hilu, 1996; Gadek et al., 1996). The gene, ˜ 1500 base pairs (bp), is located within the intron of the chloroplast gene trnK on the large single-copy section adjacent to the inverted repeat.

2.2.2 Sequence alignment and editing

Bio Edit is a biological sequence alignment editor written for Windows. An intuitive multiple document interface with convenient features makes alignment and manipulation of sequences relatively easy on computer. Several sequence manipulation and analysis options and links to external analysis programs facilitate a working environment which allows to view and manipulate sequences with simple point-and-click operations.

2.2.3 Sequence analysis and phylogenetic tree construction

Sequence analysis and phylogenetic tree construction can be done using MEGA (Molecular evolutionary genetic analysis), is an integrated tool for automatic and manual sequence alignment, inferring phylogenetic trees, mining web-based databases, estimating rates of molecular evolution, and testing evolutionary hypotheses.

3. Hypothesis

The current research work is based on the following hypothesis

1) Morphological variations among citrus varieties in Kerala would be reflected in genetic diversity among them
2) Genetic relationships among citrus varieties could be established using 'matk gene'.
3) RFLP and other bioinformatics tools would provide a better understandability and establish relationship among different morphologically identified citrus varieties.

4. Materials and Methods

4.1 Study area

Kerala state covers an area of 38,863 km^2 with a population density of 859 per km^2 and spread across 14 districts. The climate is characterized by tropical wet and dry with average annual rainfall amounts to 2,817 ± 406 mm and mean annual temperature is 26.8°C (averages from 1871-2005; Krishnakumar et al., 2009). Maximum rainfall occurs from June to September mainly due to South West Monsoon and temperatures are highest in May and November (Figure 1).

4.2 Sample collection

Sampling locations were selected in Kerala based on an elaborative baseline survey conducted during January to March 2015. The samples were collected based on an elaborative iterative survey as well as traditional knowledge from local people. Six samples were collected from different parts of Kerala, locations of the sample collection areas were recorded using a Trimble Geoexplorer II (Trimble Navigation Ltd, Sunnyvale, California) and data were transferred using GPS pathfinder Office software (Trimble Navigation Ltd, Sunnyvale, California).

Varieties of citrus are grow in south India. Selection of the plant samples was primarily based on its universal and local abundance and on its ease handling and susceptibility to the DNA isolation protocol adopted. Leaves of the plant *Citrus limon* (lemon), *Citrus maxima* or *Citrus grandis* (pommelo), *Citrus aurantifolia* (lime) were collected from Kanjirapalli, Kollappally, Athirampuzha in Kottayam district. The leaves samples were thoroughly surface sterilized in sterile distilled water to remove particulate impurities. The leaves samples were individually placed in plastic pouches and transported to lab where all samples were stored at -20°C until processed for DNA extraction.

4.3 Description of the species

Citrus aurantifolia leaves are green in colour. Leaves are 10 cm long. Adult leaves are used for DNA extraction of *Citrus aurantifolia*. *Citrus maxima* leaves are dark green in colour. They appear simple and have 12 cm long leaves which are leathery. Adult leaves are used for DNA extraction of *Citrus maxima*. *Citrus limon* leaves are light green in colour .Leaves are in oval shape. They are about 8 cm in long.

4.3.1 *Citrus aurantifolia*/lime:

Lime is a shrubby tree, to 5 m, with many thorns. The trunk rarely grows straight, with many branches that often originate quite far down on the trunk. The leaves are ovate 1–3.5 in long, resembling orange leaves. The flowers are 1 in diameter, are yellowish white with a light purple tinge on the margins. Flowers and fruit appear throughout the year but are most abundant from May to September. Lime has an odour similar to lemon, but more fresh. The juice is as sour as lemon juice, but more aromatic. The English name lime originated from Arabia limun and Persian limou.

4.3.2 *Citrus maxima*/pomelo:

Citrus maxima is a citrus tree they have 5-15 m tall, with a somewhat crooked trunk 10-30 cm thick. *Citrus maxima* branches are low, irregular and spreading. They have spines up to 5 cm long. Young branchlets are angular, often densely soft, short, hairy, and usually with Spines. Leaves are appeared in simple, and having one leaflet. Leaves are alternate, Glandular and they have 5-20 cm long, 2-12 cm wide and leathery. Petiole broadly winged to occasionally nearly wingless, up to 7cm widen. Flowers are fragrant, borne singly or in clusters of 2-10 in the leaf axils, sometimes 10-15 in terminal racemes 10-30 cm long. They have 4-5 petals Fruit are oblate or pear-shaped and10-30 cm wide. They have few seeds and seeds are large. Pomelos may flower 2-4 times a year.

4.3.3 *Citrus limon*/lemon:

Citrus limon is shrubs to medium- size trees up to about 6 m height. Trees have thin, smooth and gray- brown to greenish bark. Flowers are2-4 cm in diameter and they are fragrant. The calyx is 4-5 lobed and there are five petals. Stamens number between 20 and 40. Petals colour is white. Leaves are entire 4 to 8 cm in length. They are oval. The fruit is a hesperidium. Seeds are pale whitish to green, flattened and angular. The seed are polyembryonic.

4.3.4 Species identification and morphological characterization

a) *Citrus aurantifolia*: - Leaves are dark green in colour. They are 10cm long. The leaves are foliolate, with winged leaf stalks. The leaf lets are ovate or elliptic.

b) *Citrus maxima*: - Leaves are compound; they are 12 cm long and 3.19cm wide. They are leathery. The leaves appeared in simple and they have one leaves let. The leaves are in dark green colour.

c) *Citrus limon*:- Leaves are light green and in oval shape. They are about 4-8cm in long.

4.4 Isolation of DNA

The total genomic samples from citrus leaves were isolated using a modified cetyl trimethyl ammonium bromide (CTAB) method with the addition of β-mercaptoethanol (Shyamalamma et al., 2008). Three gram leaves samples were wiped with 70% alcohol and chopped into fine pieces and later homogenised along with 10 ml extraction buffer using a pre-chilled pestle and mortar. The extraction buffer contains 100 mM Tris-HCl, pH 8.0, 20 mM EDTA (ethylenediamine tetracetic acid), 1.4 M sodium chloride (NaCl), 3% (w/v) CTAB, 2% polyvinyl pyrrolidine (PVP) and 1% β-mercaptoethanol. The contents were slowly mixed and incubated in water bath at 65°C for 1 hr with slight shaking. The contents were brought to room temperature after incubation and 5 ml chloroform: isoamyl alcohol mixture (24:1) was added. The tubes were centrifuged at 8000 rpm for 20 min at 4°C till a clear supernatant was obtained. After the final spin, the DNA was precipitated using ice-cold isopropanol for overnight at 4°C. The tubes were further centrifuged at 5000 rpm and the pellet washed with 2-3 drops of 70% alcohol, air dried and redissolved in 50 µl Tris-EDTA (TE) buffer and stored at -20°C till further analysis (Simon et al., 2007; Shyamalamma et al., 2008; Xiaoming and Xiuxin, 2001; Allen et al., 2006).

4.5 Quantification of DNA

The purity of the isolated DNA was checked spectrophometrically in a UV-Vis spectrophotometer by checking the 260/280 values (Thompson and Dvorak, 1989; Muller et al., 2003). Concentration of DNA (ng/µl) were calculated using the formula

DNA (ng/µl) = OD @ A260 x 50 x 100 x 0.1

Where OD @ A260 is the optical density at absorbance 260 nm

50 is the calculation factor

100 is the dilution factor

0.1 is the total volume of DNA

4.6 PCR primers

Forward and reverse primers for rbcL and matK genes were custom synthesized according to the primer pair sequences obtained from the earlier work of Bafeel et al. (2011) on several plant species using rbcL and matK gene specific primer pairs viz; RbcL-1and RbcL-724R for rbcL gene and MatK-390F and MatK-1326R for matK gene.

Forward primer for rbcL gene	RbcL-1 (5′ATGTCACCACAAACAGAAAC 3′)
Reverse primer for rbcL gene	RbcL-724R (5′TCGCATGTACCTGCAGTAGC 3′)
Forward primer for matK gene	MatK-390F (5′CGATCTATTCATTCAATATTTC3′)
Reverse primer for matK gene	MatK-1326R (5′TCTAGCACACGAAAGTCGAAGT3′)

4.7 PCR amplification

The polymerase enzymes, adaptors and primers were purchased from Genie life science technology, Bangalore, India. The PCR reaction was performed with a Biorad MJPt100 thermocycler (Bio-Rad Laboratories, Bangalore, India). PCR amplifications were performed on two stages. The pre-selective amplification was performed with an amplification profile of 94°C for 30 s, anneling at 56°C for 1 min, extension at 72°C for 1 min, repeated for 20 cycles, and then cooling at 10°C for 30 min. Further amplification was performed with a cycling profile of 94°C for 30 s, 65°C for 30 s, 72°C for 24 cycles followed by a cooling of 10°C for 30 min. The primers used for the selective amplification ended with three nucleotide extensions at 3' ends. The PCR products were stored at -20°C in a deep-freezer. Electrophoresis of the samples was carried out on agarose gels, by loading 10 µl of each DNA samples at 50 V for 3 hrs tracking dye has properly moved across the gel. The gels were later viewed on a gel docking station and photographed.

4.8 Data sequencing

The PCR products were cleaned up using GenElute[TM] PCR Clean-Up Kit (Sigma-Aldrich). Purified PCR products were sequenced by dideoxy chain termination method (Sanger et al., 1977) using AB3730XL capillary sequencer for the six jackfruit samples.

4.9 Sequence alignment

Multiple alignments of sequences were performed using CLUSTAL X version 2 (Larkin et al., 2007) alignment editor. Alignment was then manually checked and corrected. Nucleotide sequence characteristics after alignment were analysed using the program DnaSP version 4.10 (Rozas et al., 2003).

4.10 Phylogenetic and molecular evolutionary analysis

Phylogenetic and molecular evolutionary analysis was conducted using MEGA version 4 (Tamura et al., 2007). Sequence data was subsequently analysed using distance (Neighbour-Joining) and Maximum Parsimony methods. The sampling error of Neighbour-Joining (NJ) and Maximum Parsimony Trees was analysed using bootstraps of 10,000 replicates where possible followed by the construction of Majority Rule Trees. Pair wise sequence divergence among populations was calculated according to Kimura two-parameter model (Kimura, 1980). The number and rate of transitions/transversions were also calculated using the program MEGA.

4.11 Statistical analysis

The survey results were analyzed and descriptive statistics were done using SPSS 12.0 (SPSS Inc., an IBM Company, Chicago, USA) and graphs were generated using Sigma Plot 7 (Systat Software Inc., Chicago, USA).

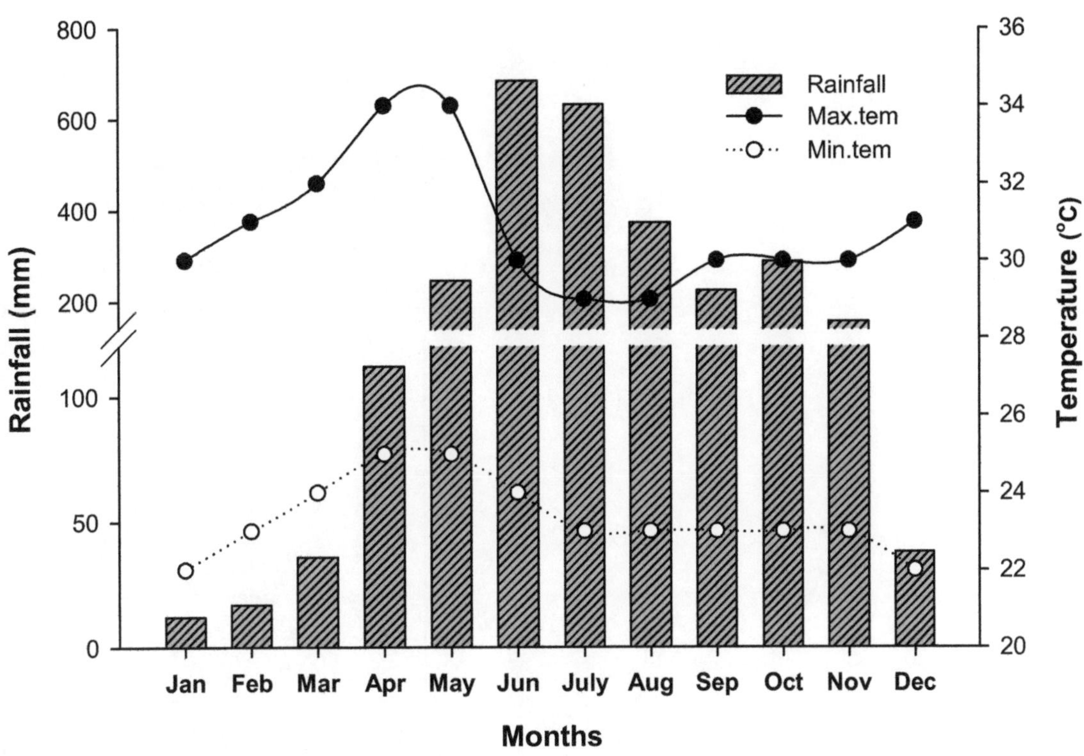

Figure 1. Mean monthly rainfall (mm), maximum and minimum temperatures (°C) in Kerala, India (1871-2005; Krishnakumar et al., 2009).

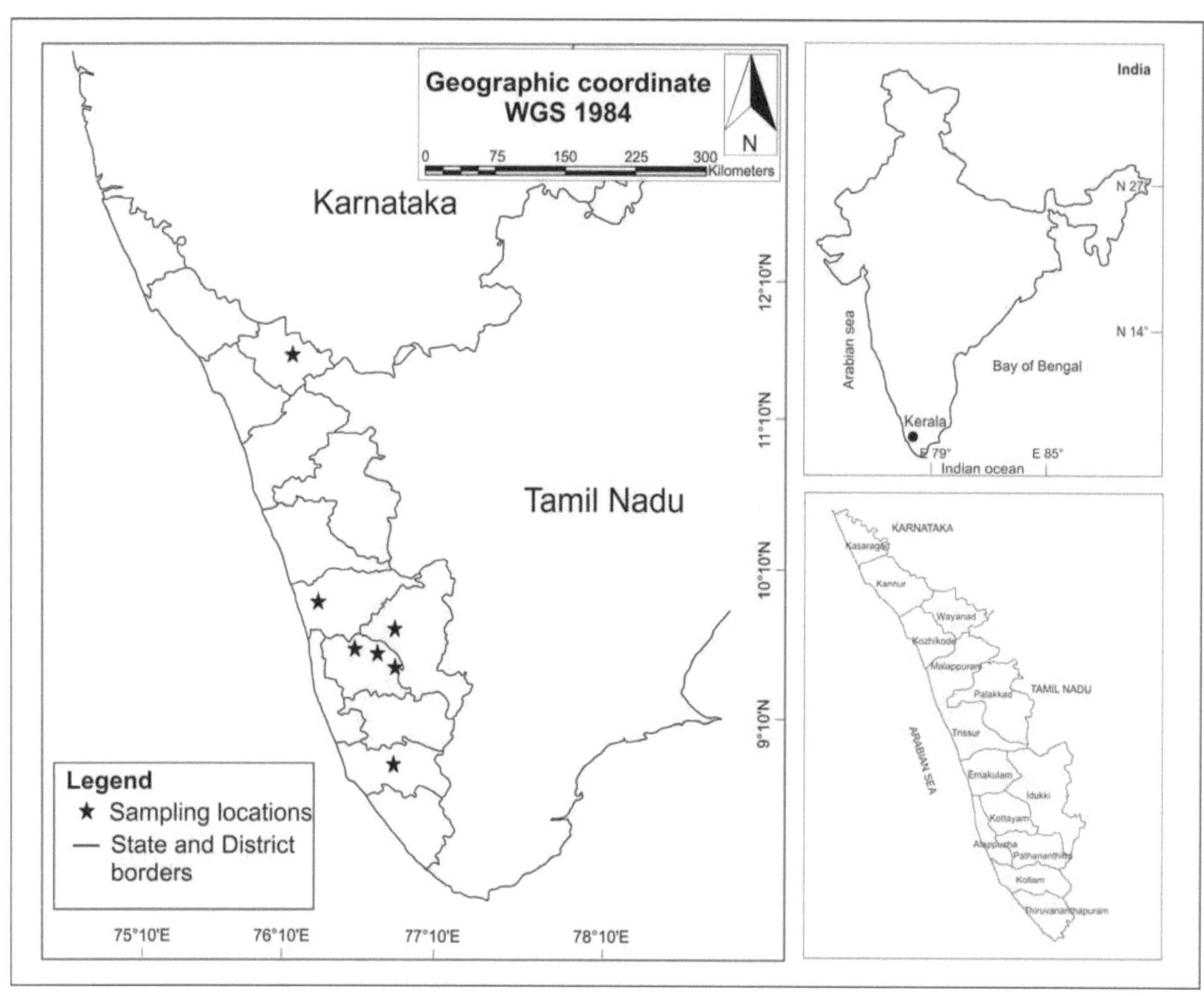

Figure 2. Map of Kerala showing the various sample collection points. Authors own work.

Figure 3. Description of the various citrus varieties a), e) *Citrus aurantiifoli* tree bearing fruits, b) *Citrus maxima*, c) *Citrus limon*, d) orange fruit and blossom, f) pompia. Photo courtesy: Wikipedia.

Figure 4. Description of the various citrus varieties a) Lime fruits, b) amber sweet oranges, c) Clementine peeled, d) cross section of the various types of citrus fruits, e) citrus canker on fruit, f) odichukuthi lime cross section. Photo courtesy: Wikipedia.

Figure 5. Description of the various citrus varieties a) citrus fruit cut open, b) *Citrus australasica* fruit, c) citrus leaf, d) unripe mandarin fruit, e) odichukuthi lime. Photo courtesy: Wikipedia.

Figure 6. Key lime (*Citrus aurantifolia*) description a) ripe key lime fruit, b) mature key lime fruit, c) and d) key lime flowers. Photo courtesy: Wikipedia.

Figure 7. Pomelo (*Citrus maxima/Citrus grandis*) description a) mature *Citrus maxima* fruit, b) pomelo fruit cut open, c) *Citrus maxima* flower, d) *Citrus maxima* leaf, e) and f) pomelo flesh in red and cream colour. Photo courtesy: Wikipedia.

Figure 8. Lemon description a) lemon fruit and flower, b) lemon fruit on tree, c) leamon flower, d) lemon fruit cut open, e) lemon fruits on tree. Photo courtesy: Wikipedia.
.

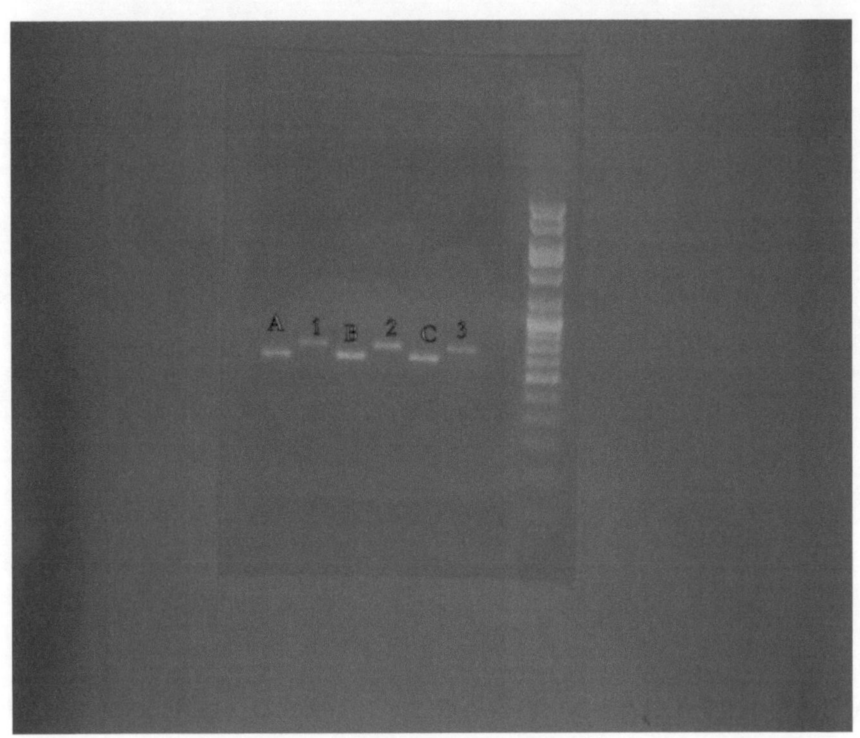

Figure 9. Agarose gel electrophoresis of PCR product; A, B and C are PCR products of rbcL gene and 1, 2 and 3 are PCR products of matK gene. Authors own image.

Figure 10. BLAST result of rbcL gene of *Citrus aurantifolia* 1. Authors own work.

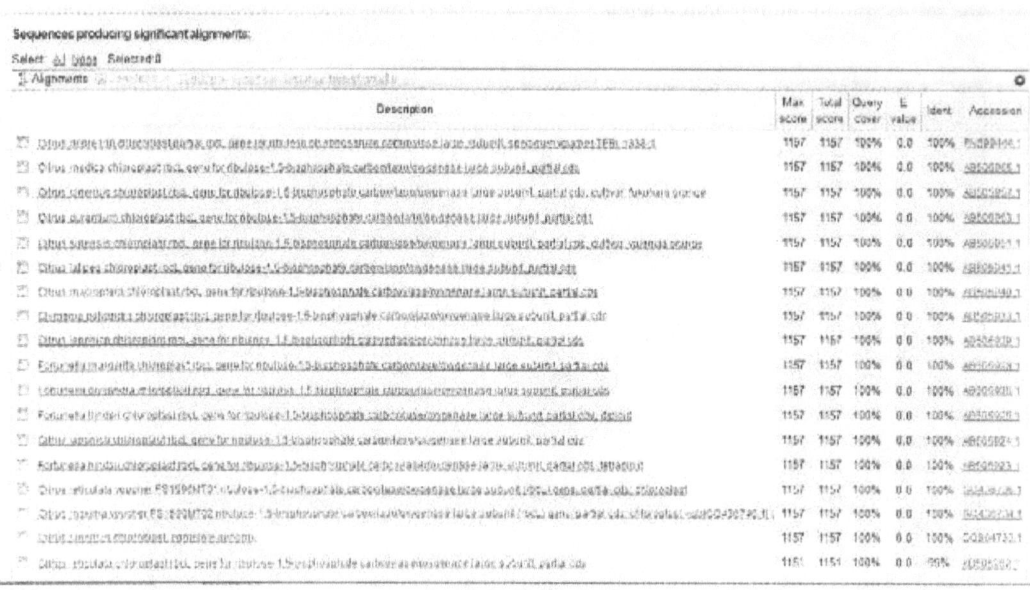

Figure 11. BLAST result of rbcL gene of *Citrus aurantifolia* 2. Authors own work.

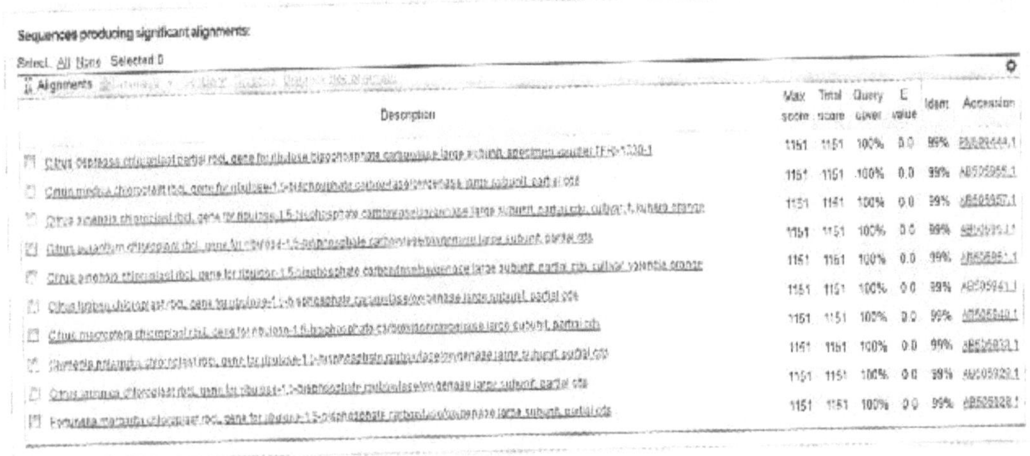

Figure 12. BLAST result of rbcL gene of *Citrus aurantifolia* 3. Authors own work.

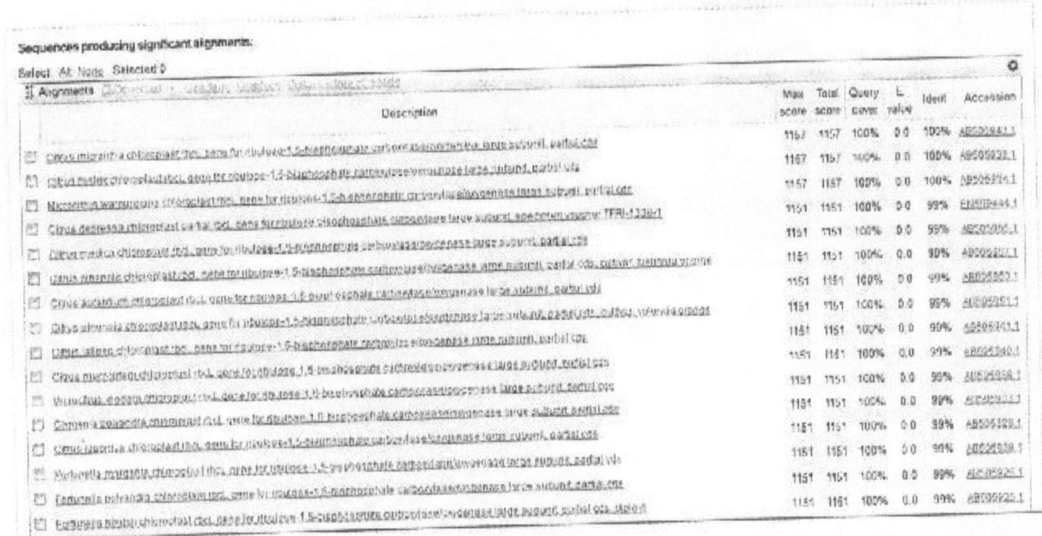

Figure 13. BLAST result of rbcL gene of *Citrus lemon* 1. Authors own work.

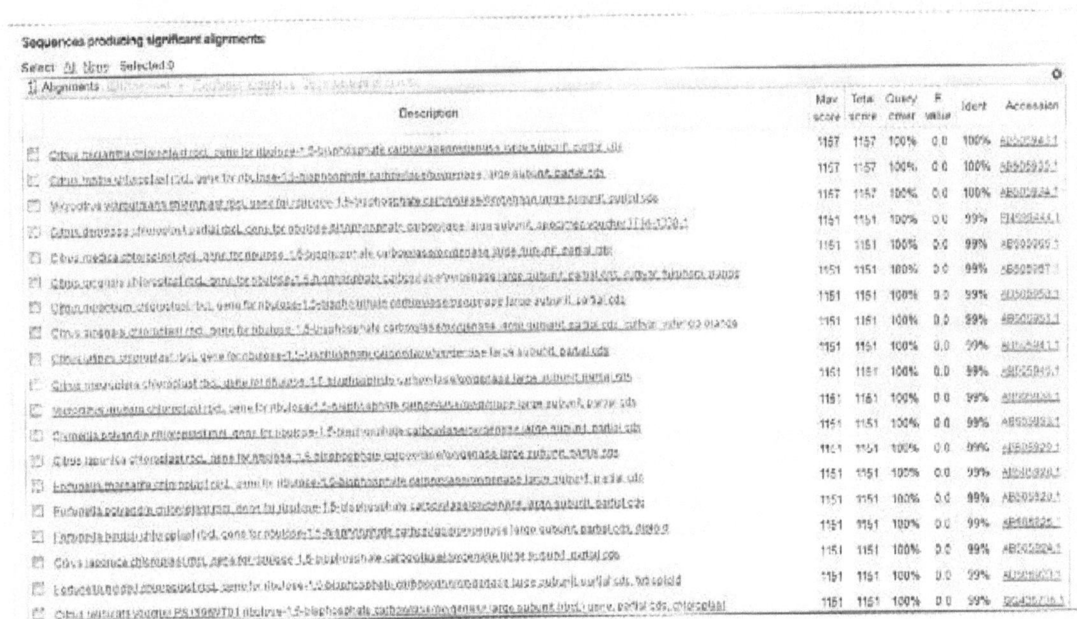

Figure 14. BLAST result of rbcL gene of *Citrus lemon* 2. Authors own work.

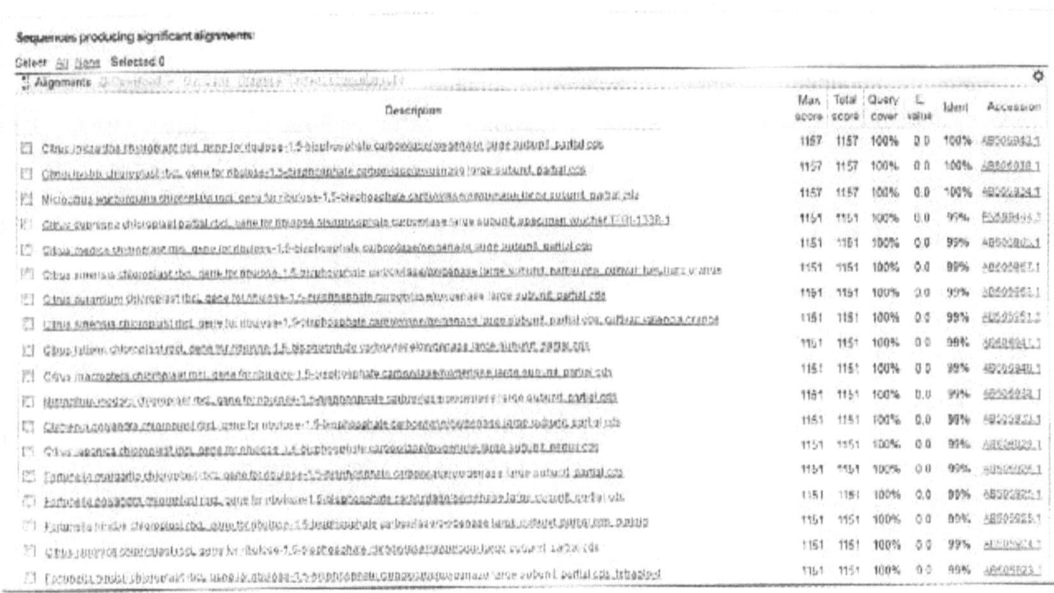

Figure 15. BLAST result of rbcL gene of *Citrus lemon* 3. Authors own work.

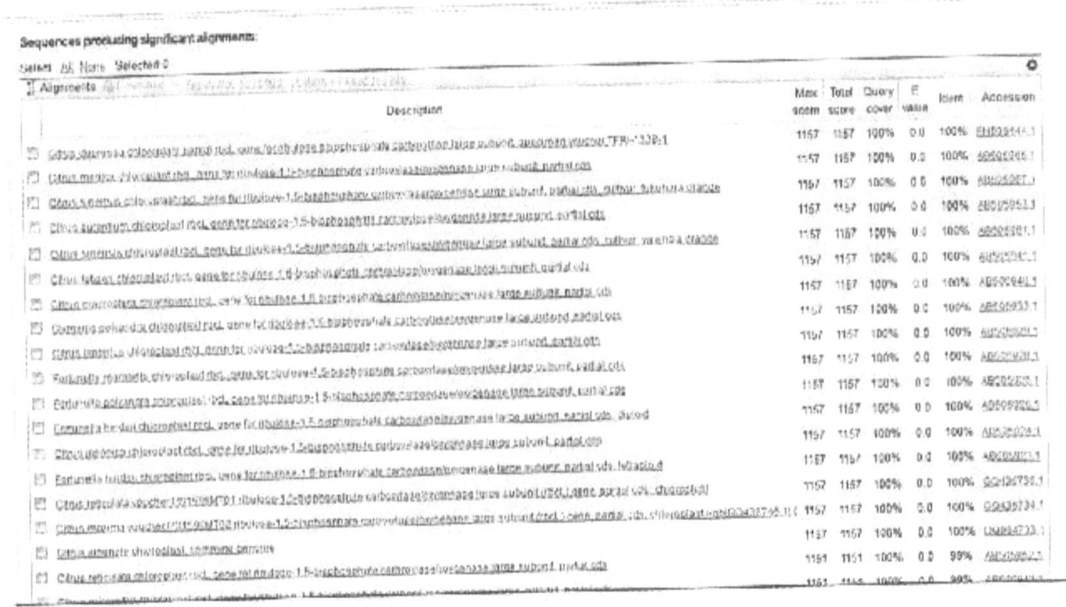

Figure 16. BLAST result of rbcL gene of *Citrus maxima* 1. Authors own work.

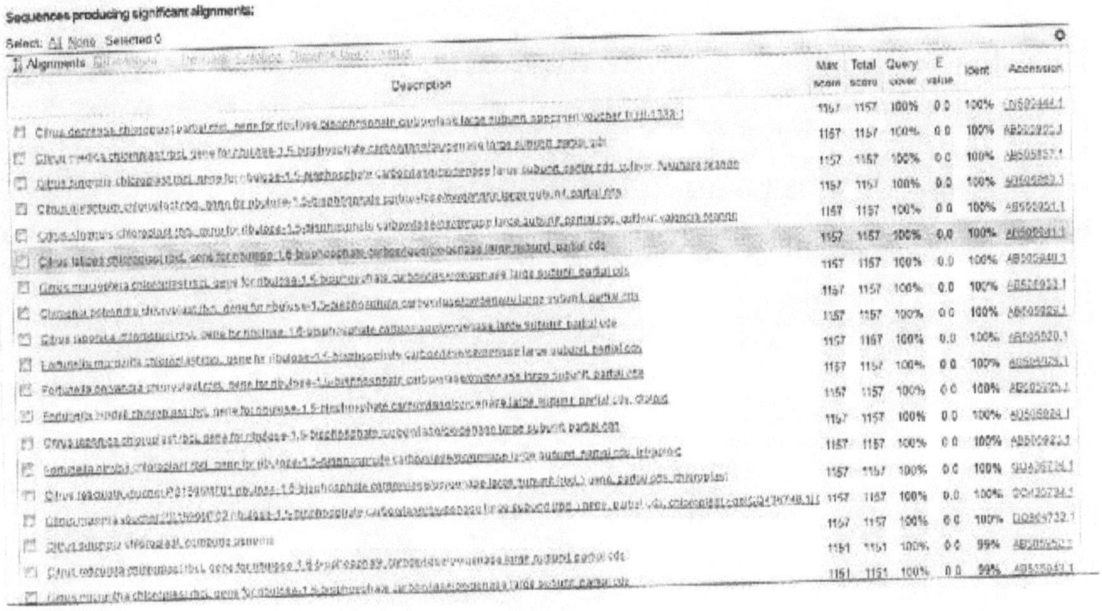

Figure 17. BLAST result of rbcL gene of *Citrus maxima* 2. Authors own work.

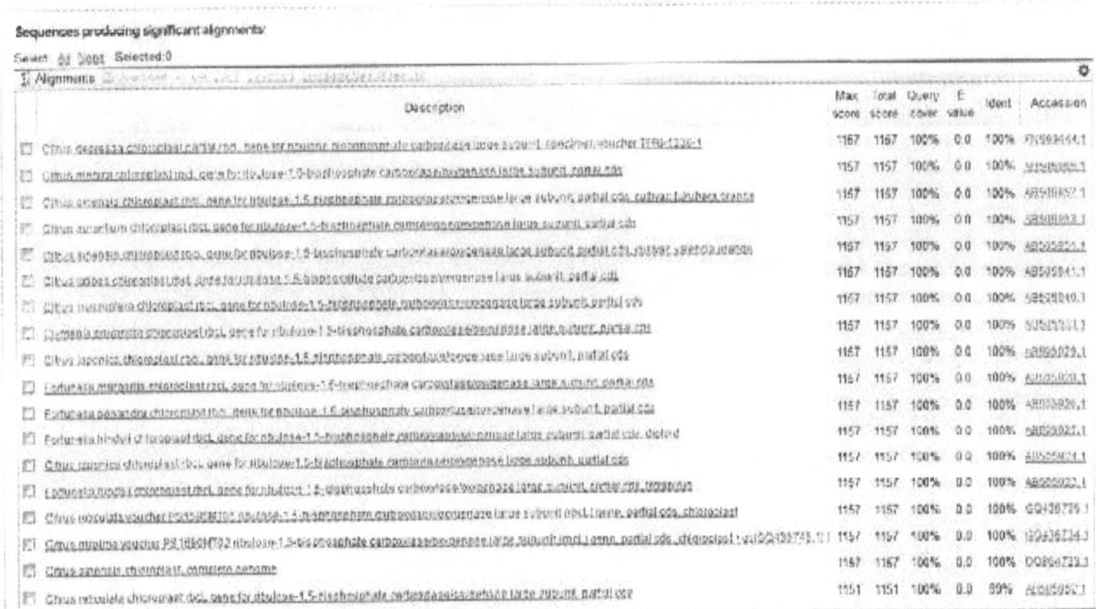

Figure 18. BLAST result of rbcL gene of *Citrus maxima* 3. Authors own work.

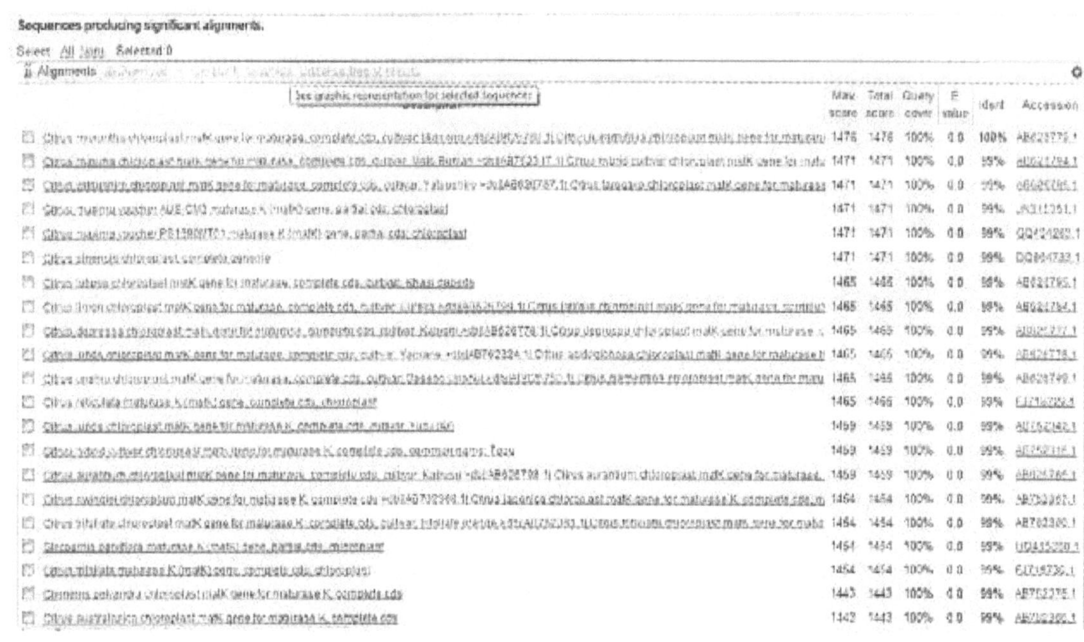

Figure 19. BLAST result of matK gene of *Citrus aurantifolia* 1. Authors own work.

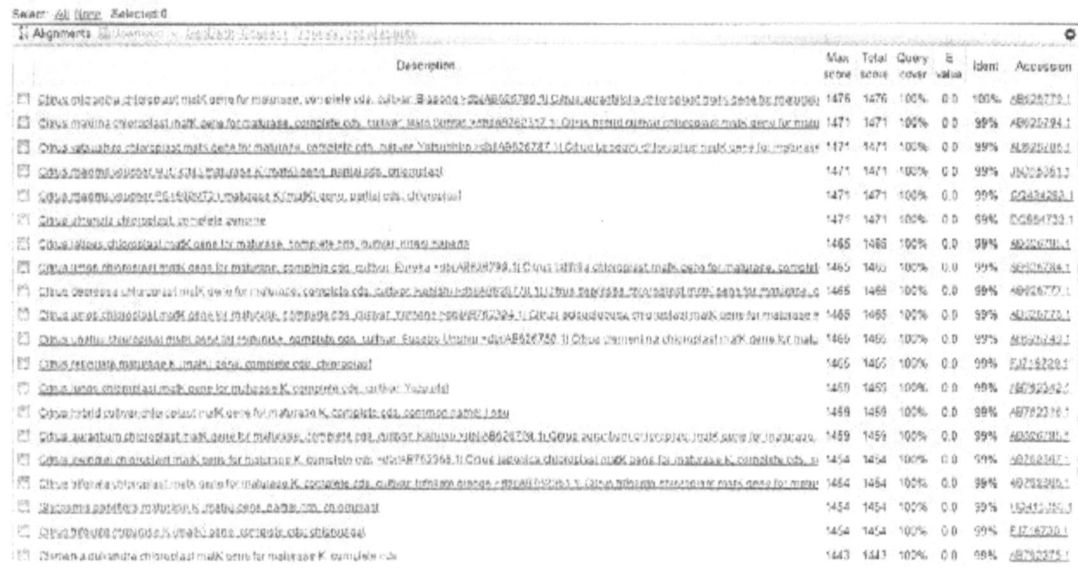

Figure 20. BLAST result of matK gene of *Citrus aurantifolia* 2. Authors own work.

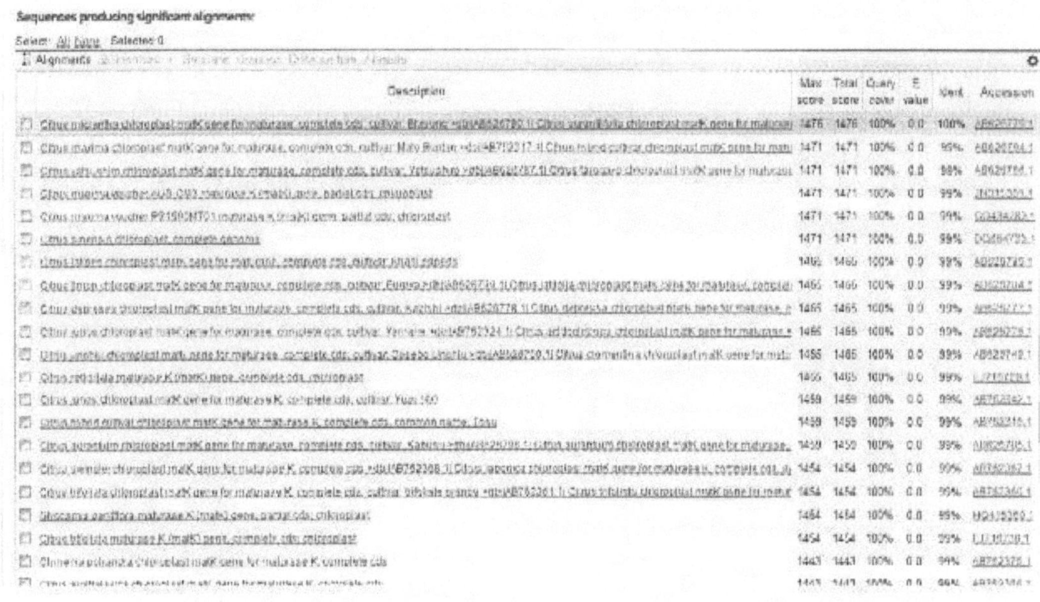

Figure 21. BLAST result of matK gene of *Citrus aurantifolia* 3. Authors own work.

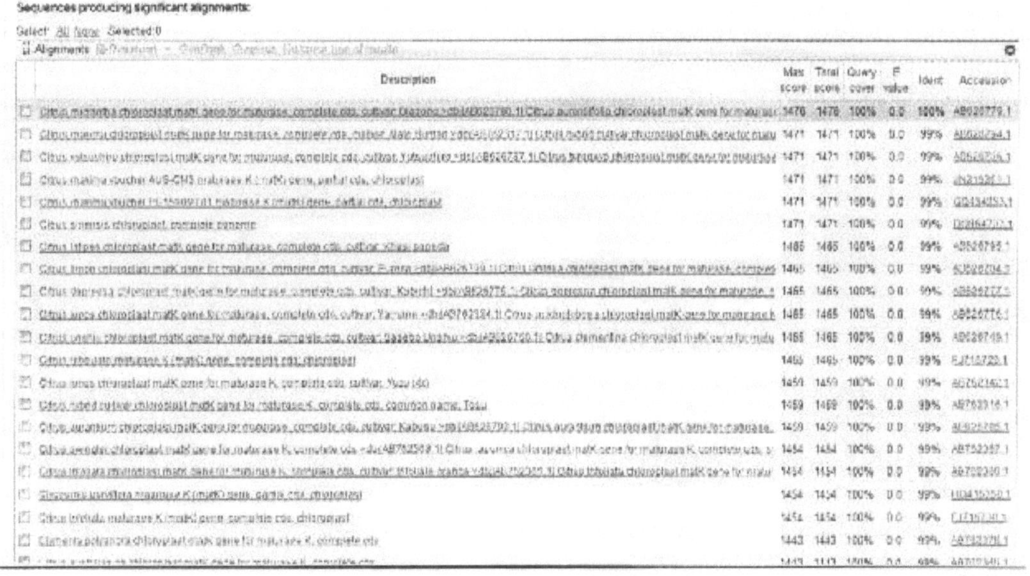

Figure 22. BLAST result of matK gene of *Citrus lemon* 1. Authors own work.

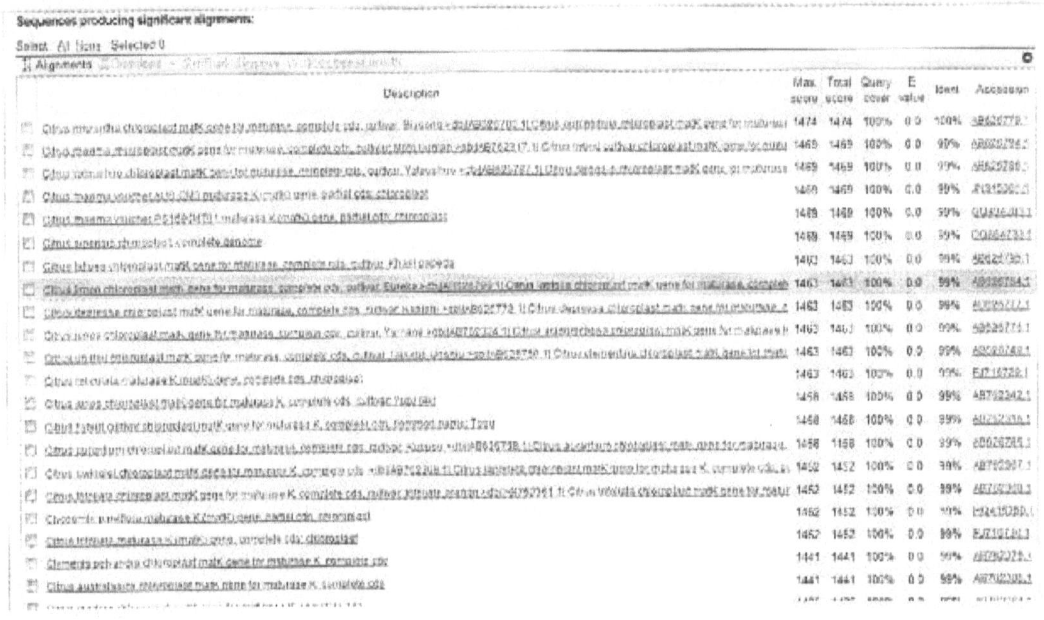

Figure 23. BLAST result of matK gene of *Citrus lemon* 2. Authors own work.

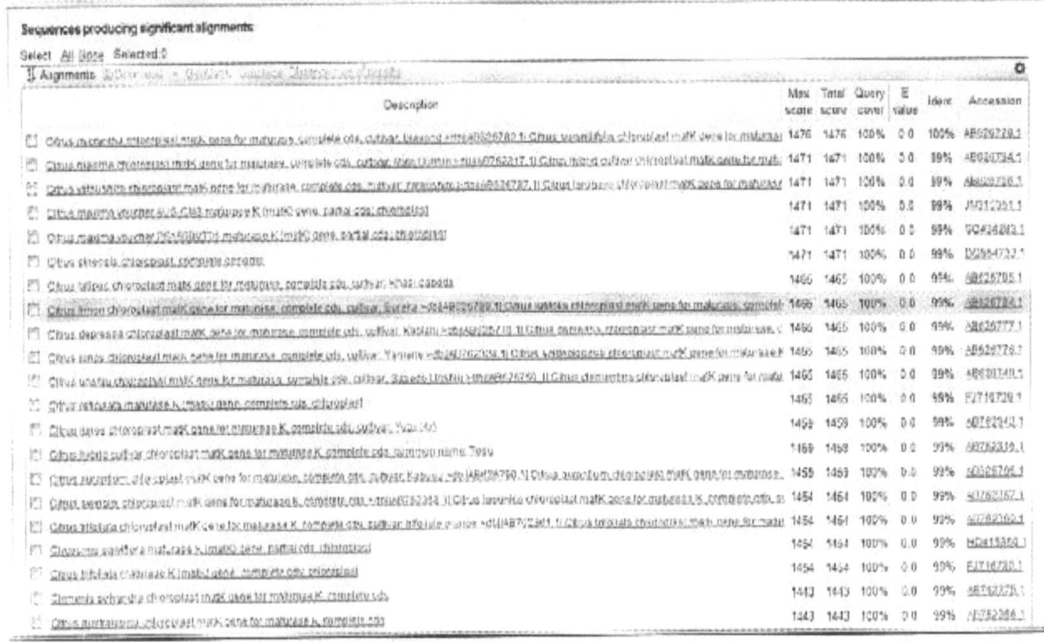

Figure 24. BLAST result of matK gene of *Citrus lemon* 3. Authors own work.

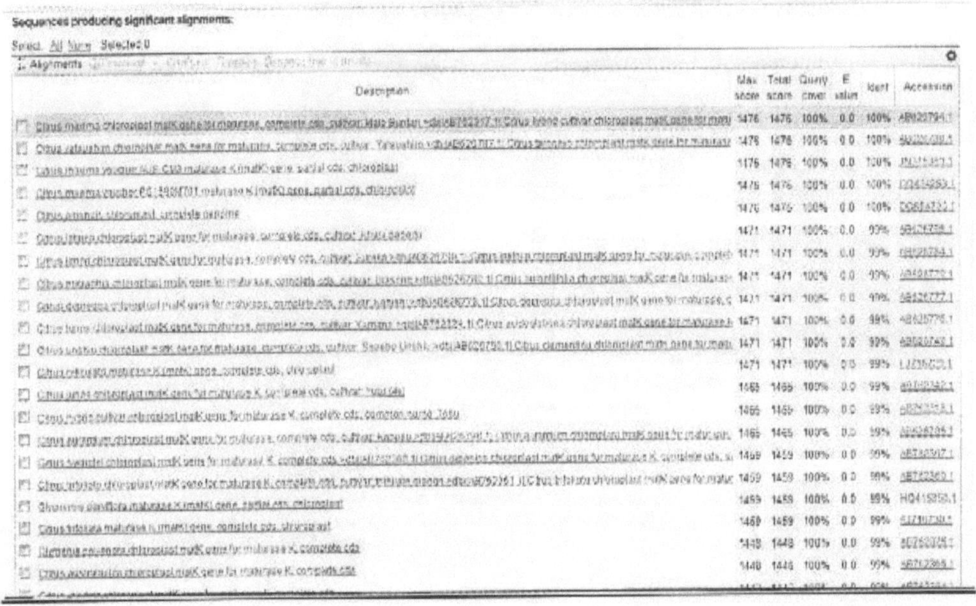

Figure 25. BLAST result of matK gene of *Citrus maxima* 1. Authors own work.

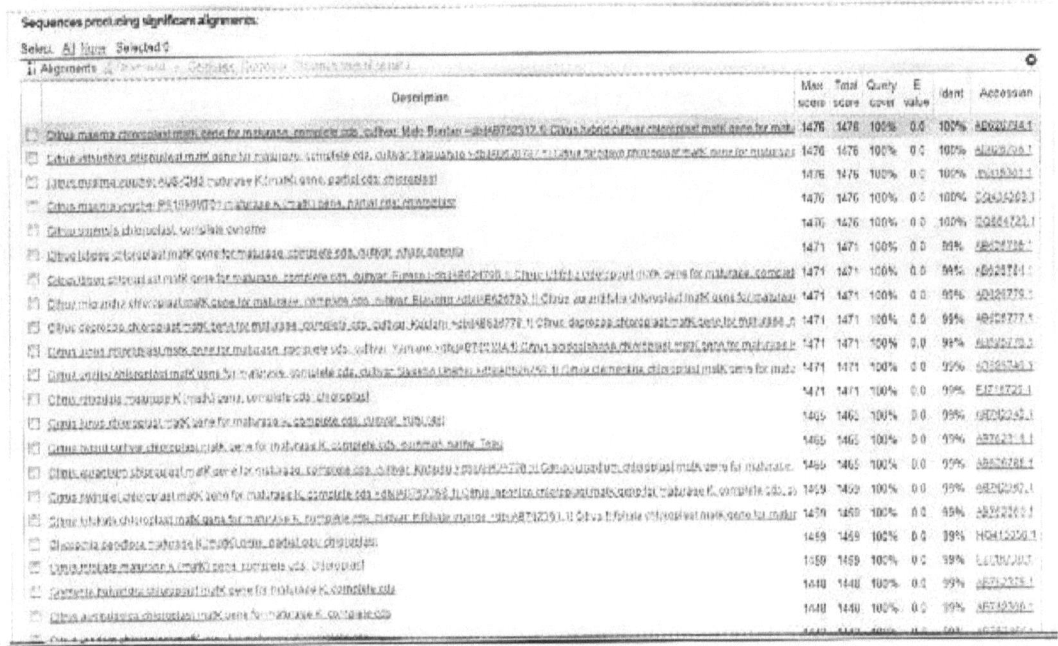

Figure 26. BLAST result of matK gene of *Citrus maxima* 2. Authors own work.

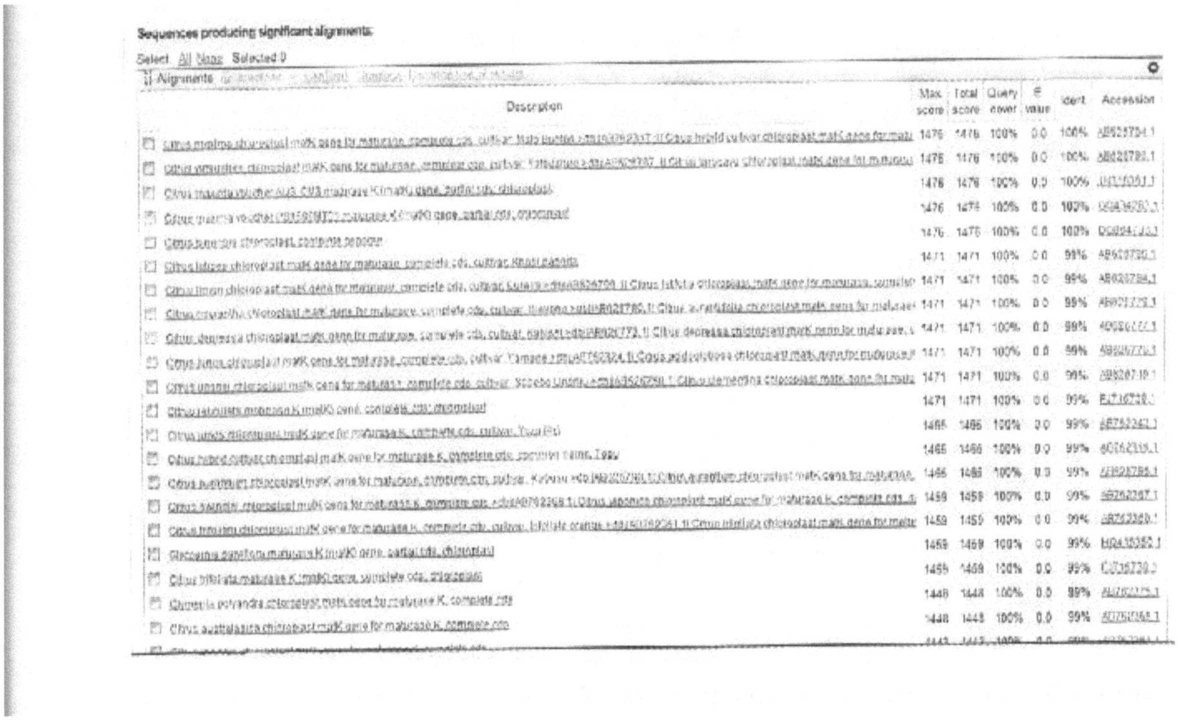

Figure 27. BLAST result of matK gene of *Citrus maxima* 3. Authors own work.

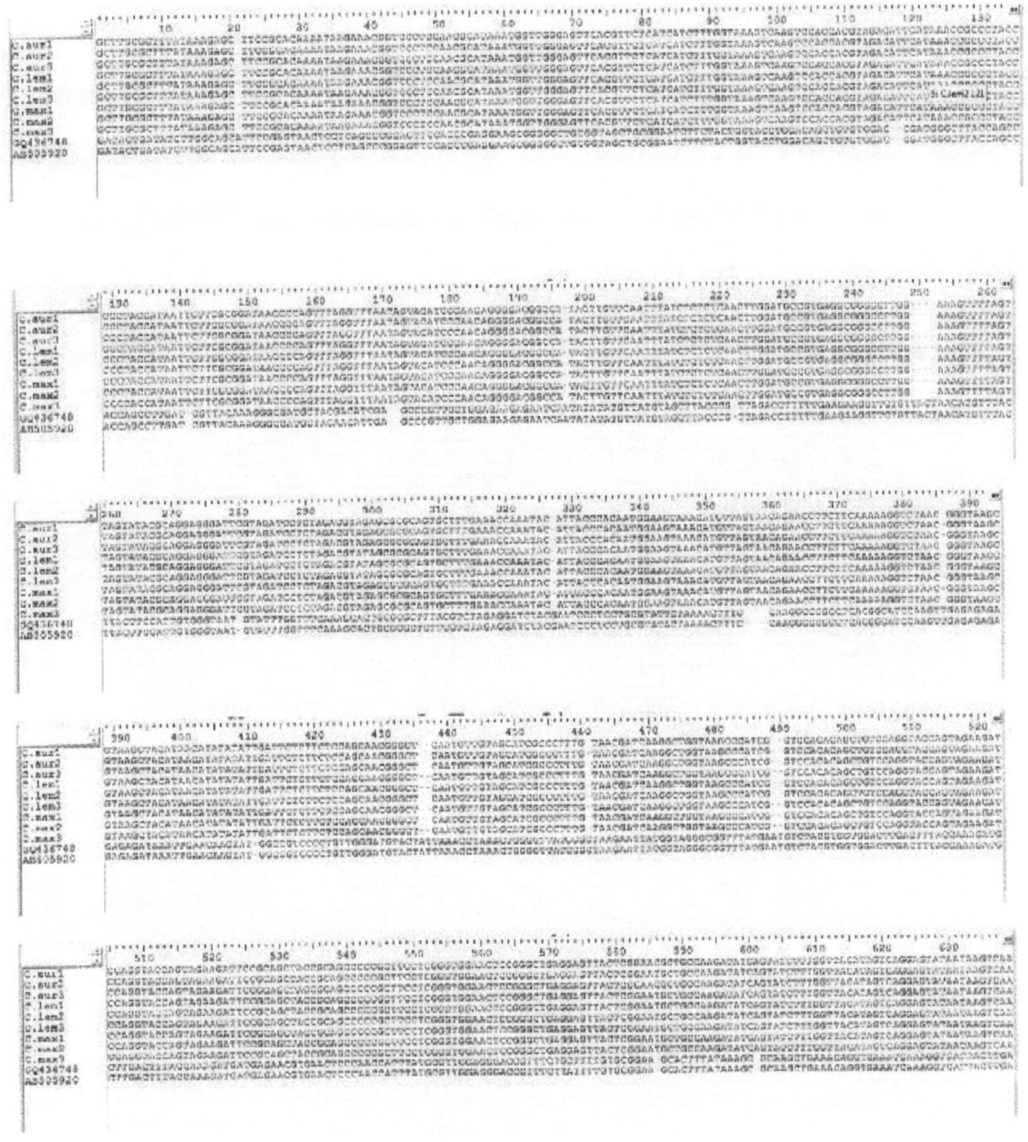

Figure 28. Multiple sequence alignment of rbcL gene of Citrus species. Authors own work.

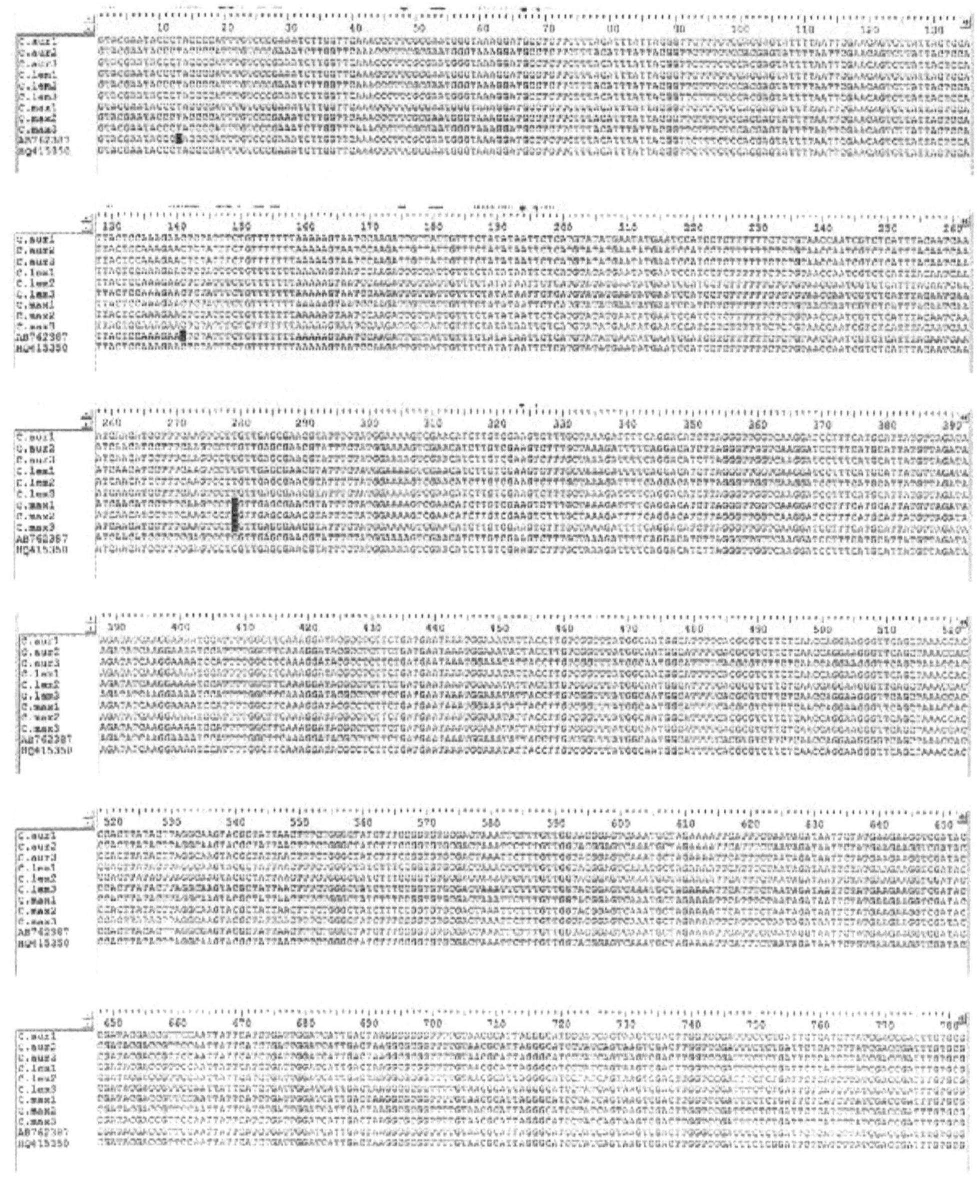

Figure 29. Sequence alignment of matK gene of Citrus species. Authors own work.

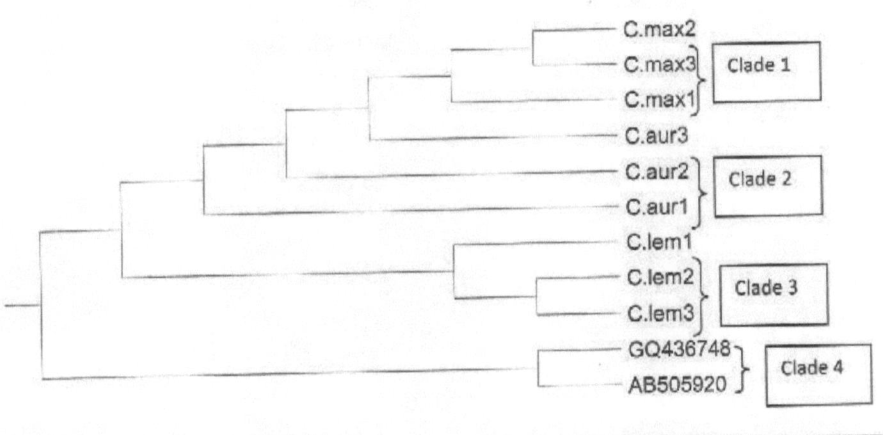

Figure 30. Phylogenetic tree based on partial sequence rbcL gene of Citrus species by UPGMA method. Authors own work.

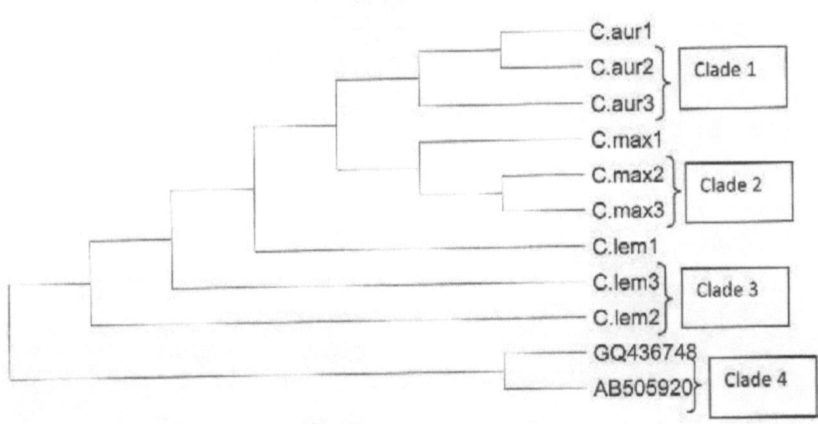

Figure 31. Phylogenetic tree based on partial sequence rbcL gene of Citrus species by Maximum likelihood method. Authors own work.

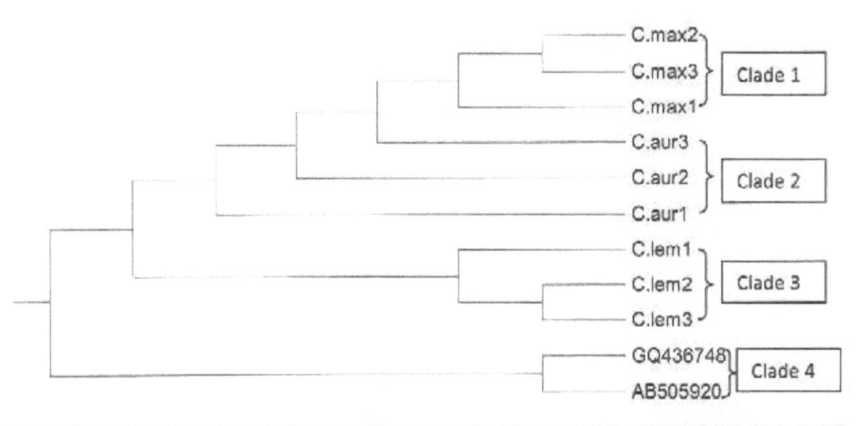

Figure 32. Phylogenetic tree based on partial sequence matK gene of Citrus species by UPGMA method. Authors own work.

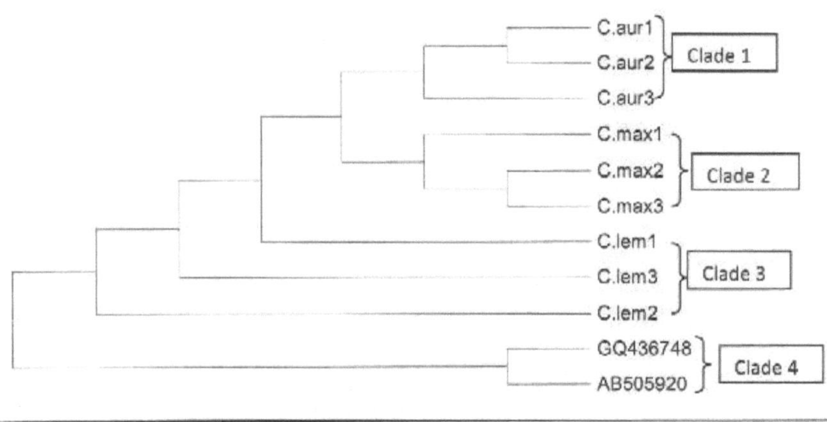

Figure 33. Phylogenetic tree based on partial sequence rbcL gene of Citrus species by Maximum likelihood method. Authors own work.

Table 1. Details of the Citrus varieties collected from Kerala.

| Name of variety | Details | | | Seed |
	Taluk	Place	GPS position	Features
Citrus aurantifolia	Menachil	Kanjirapapplly	9°33'26.17" N 76°47'21.96" E	The leaves are foliolate with winged leaf stalks, Leaflets ovate or elliptic, leaves dark green in colour with 10 cm long
Citrus maxima	Menachil	Kollappally	9°45'47.46" N 76°42'1.34" E	Compound leaves with 12 cm long and 3.19 cm wide, dark green colour in simple pattern
Citrus limon	Kottayam	Athiirampuzha	9°40'4.44" N 76°31'58.62" E	Leaves 4 to 8 cm long with light green colour and oval in shape

Table 2. Different vernacular names of *Citrus aurantifolia* (Key lime) in India.

Language	**Names**
Scientific names	*Citrus aurantifolia*
Name in various Indian languages	
Sanskrit	Matulunga, Nimbukah
Hindi	Kaghzi-nimbu, Nimbu
Bengali	Lebu
Marathi	Ambatanimbu, Limbu, Mavalanga
Kannada	-
Konkani	-
Malayalam	-
Tamil	Champalam, Champiram
Telugu	Nimma

Table 3. Different vernacular names of *Citrus maxima* (Pomelo) in India.

Language	Names
Scientific names	*Citrus maxima*
	Citrus grandis
Name in various Indian languages	
Sanskrit	Karuna, Mallikapuspa
Hindi	Batawi-nimbu, Cakotara, Papar-mas
Bengali	Batabi lebu
Marathi	Bampara, Cakotara, Papannasa
Kannada	Cakota hannu
Konkani	Toranda
Malayalam	Kampilinarannga
Tamil	Metukku, Pampalimacu, Pommacu
Telugu	Pamparapanasa

Table 4. Different vernacular names of *Citrus limon* (Lemon) in India.

Language	Names
Scientific names	*Citrus limon*
Name in various Indian languages	
Sanskrit	Karuna, Mallikapuspa
Hindi	Nimbu
Bengali	-
Marathi	-
Kannada	-
Konkani	-
Malayalam	-
Tamil	Elumicchai
Telugu	-

Table 5. Pair wise phylogenetic distance matrix based on partial rbcL gene Citrus species. The overall mean distance of selected sample is 0.001.

		1	2	3	4	5	6	7	8	9	10	11
1	C.aur1											
2	C.aur2	0.000										
3	C.aur3	0.000	0.000									
4	C.lem1	0.002	0.002	0.002								
5	C.lem2	0.002	0.002	0.002	0.000							
6	C.lem3	0.002	0.002	0.002	0.000	0.000						
7	C.max1	0.000	0.000	0.000	0.002	0.002	0.002					
8	C.max2	0.000	0.000	0.000	0.002	0.002	0.002	0.000				
9	C.max3	0.000	0.000	0.000	0.002	0.002	0.002	0.000	0.000			
10	GQ436748	1.252	1.252	1.252	1.256	1.256	1.256	1.252	1.252	1.252		
11	AB505920	1.240	1.240	1.240	1.244	1.244	1.244	1.240	1.240	1.240	0.012	

Table 6: Pairwise Phylogenetic distance matrix based on partial *rbcL* gene of *Citrus* species

Table 6. Pair wise phylogenetic distance matrix based on matK gene Citrus species. The overall mean distance of selected sample is 0.001.

		1	2	3	4	5	6	7	8	9	10	11
1	C.aur1											
2	C.aur2	0.000										
3	C.aur3	0.000	0.000									
4	C.lem1	0.002	0.002	0.002								
5	C.lem2	0.002	0.002	0.002	0.000							
6	C.lem3	0.002	0.002	0.002	0.000	0.000						
7	C.max1	0.000	0.000	0.000	0.002	0.002	0.002					
8	C.max2	0.000	0.000	0.000	0.002	0.002	0.002	0.000				
9	C.max3	0.000	0.000	0.000	0.002	0.002	0.002	0.000	0.000			
10	GQ436748	1.252	1.252	1.252	1.256	1.256	1.256	1.252	1.252	1.252		
11	AB505920	1.240	1.240	1.240	1.244	1.244	1.244	1.240	1.240	1.240	0.012	

Table 7: Pairwise Phylogenetic distance matrix based on maturaseK gene of *Citrus* species

5. Results and discussion

5.1 Blast search for rbcL gene of Citrus species

The similarity search of three varieties of Citrus species analysis by basic local alignment search tool (BLAST) programme of NCBI (National center for biotechnology information). The BLAST search results showed all the three samples of *Citrus aurantifolia* is having 99% similarity to *Citrus depressa* and *Citrus medica*. The three samples of *Citrus lemon* showed 100% similarity to *Citrus micrantha*. The three samples of *Citrus maxima* showed 100% similarity to *Citrus depressa* and *Citrus medica*.

5.2 Blast search for matK gene of Citrus species

The similarity search of amplified regin of matK gene of the three varieties of Citrus species were analyzed using BLAST programme. The BLAST search showed that the three samples of *Citrus aurantifolia* and *Citrus lemon* showed 100% similarity to *Citrus micrantha* and *Citrus maxima*.

5.3 Sequence alignment
5.3.1 Multiple sequence alignment of partial rbcL gene sequence of Citrus species

The multiple sequence alignment of rbcL gene of Citrus species were done by Bio Edit software. Nine sequences belonging to the three Citrus species were subjected to sequence alignment along with partial rbcL gene sequence of *Glycosmis pentaphylla* (Acc.No. GQ436748) and *Atalantia monophylla* (Acc.No. AB505920) were used as outgroups.

5.3.2 Multiple sequence alignment of partial matK gene sequence of Citrus species

The multiple sequence alignment of matK gene of Citrus species were done by Bio Edit software. Nine sequences belonging to the three Citrus species were subjected to sequence alignment along with partial matK gene sequence of *Paramignya lobata* (Acc.No. AB762387) and *Glycosmis parviflora* (Acc.No. HQ415350) were used as out groups.

5.4 Construction of phylogentic tree
5.4.1 Phylogentic tree using rbcL gene

The cladogram has four distinct clades. Top most clade consisting of *Citrus maxima* 3, *Citrus maxima* 1. Next clade consisting of *Citrus aurantifolia* 2 and *Citrus aurantifolia* 1. The next clade consisting of *Citrus lemon* 2 and *Citrus lemon* 3. The fourth and final clade has only two species; the outgroups *Atalantia monophylla* and *Glycosmis pentaphylla*. Similar results were obtained using maximum likelihood method.

5.3 Phylogenetic tree using matk gene

The cladogram consist of three distinct clades; top most clade 1 consist of *Citrus maxima* 2, *Citrus maxima* 3 and *Citrus maxima* 1. Clade 2 consist of *Citrus aurantifolia* 3, *Citrus aurantifolia* 2 and *Citrus aurantifolia* 1, while the clade 3 consist of *Citrus lemon* 1, *Citrus lemon* 2 and *Citrus lemon* 3. The fourth and last clade consist of two species; the out groups *Atalantia monophylla* and *Glycosmis pentaphylla*. In each group, there exists high similarity between species within the group. Similar results were obtained using maximum likelihood method.

5.4 Calculation of the genetic distance
5.4.1 Genetic distance using rbcL gene

The genetic distance between three samples of *Citrus aurantifolia* were same. The genetic distance between three samples of *Citrus aurantifolia* (1, 2 and 3) and *Citrus lemon* (1, 2 and 3) is 0.002. The genetic distance between three samples of *Citrus lemon* (1, 2 and 3) and three samples of *Citrus maxima* (1, 2 and 3) were 0.002.

5.4.2 Genetic distance using matK gene

The genetic distance between three samples of *Citrus aurantifolia* were same. The genetic distance between three samples of *Citrus aurantifolia* (1, 2 and 3) and *Citrus lemon* (1, 2 and 3) is 0.002. The genetic distance between three samples of *Citrus lemon* (1, 2 and 3) and three samples of *Citrus maxima* (1, 2 and 3) were 0.002. The genetic distance between three samples of *Citrus aurantifolia* (1, 2 and 3) and three samples of *Citrus maxima* (1, 2 and 3) were same. The genetic distance between the three samples of *Citrus maxima* were also same.

6. Conclusions

Being ultimately, phylogenetic nomenclature is a result of Darwin's discovery that the diversity and history of life is best represented in tree-shaped diagrams. This discovery immediately led to changes in the existing classifications. In 1866, the controversial biologist Ernst Haeckel for the first time reconstructed a single tree of all life and immediately proceeded to translate it into a classification. This classification was rank-based, in accordance with the only code of biological nomenclature that existed at the time. The origin of phylogenetic study can be dated to 1986, when Jacques Gauthier used phylogenetic definitions for the first time in a published work. Molecular phylogenetics is the analysis of hereditary molecular differences, mainly in DNA sequences, to gain information on an organism's evolutionary relationships. The result of a molecular phylogenetic analysis is expressed in a phylogenetic tree. Molecular phylogenetics is one aspect of molecular systematics, a broader term that also includes the use of molecular data in taxonomy and biogeography. The theoretical frameworks for molecular systematics were laid in the 1960s in the works of Emile Zuckerkandl, Emanuel Margoliash, Linus Pauling, and Walter M. Fitch. Applications of molecular phylogeny were pioneered by Charles G. Sibley, Herbert C. Dessauer, and Morris Goodman , followed by Allan C. Wilson, Robert K. Selander, and John C. Avise. Work with electrophoresis began around 1956. In the period of 1974–1986, DNA-DNA hybridization was the dominant technique. Molecular phylogenies can be affected by myriad problems, including long-branch attraction, saturation, and taxon sampling problems: This means that strikingly different results can be obtained by applying different models to the same dataset.

Acknowledgements

The authors are grateful for the cooperation of the management of Mar Augusthinose college for necessary support. This research work will not be possible with the technical support from the Marine Biotechnology Division, Central Marine Fisheries Research Institute (CMFRI), Kochi. Technical assistance from Binoy A Mulanthra is also acknowledged.

References

Algabal, A.Q.A.Y., Papanna, N., & Simon, L. (2011). Estimation of Genetic Variability in Tamarind (*Tamarindus indica* L.) using RAPD Markers. *International Journal of Plant Breeding*, 5(1), 10-16.

Aljanabi, S.M., & Martinez, I. (1997). Universal and rapid salt-extraction of high quality genomic DNA for PCR-based techniques. *Nucleic Acids Research*, 25(22), 4692-4693.

Allen, G.C., Flores-Vergara, M.A., Krasynanski, S., Kumar, S., & Thompson, W.F. (2006). A modified protocol for rapid DNA isolation from plant tissues using cetyltrimethylammonium bromide. *Nature Protocols*, 1(5), 2320-2325.

Barkley, N. A., Roose, M. L., Krueger, R. R., & Federici, C. T. (2006). Assessing genetic diversity and population structure in a citrus germplasm collection utilizing simple sequence repeat markers (SSRs). *Theoretical and Applied Genetics*, *112*(8), 1519-1531.

Bruneton, J. (1999). Toxic plants dangerous to humans and animals. Intercept Limited.

Burkill, H. M. (1997). *The useful plants of west tropical Africa, Vols. 1-3* (No. 2. ed.). Royal Botanic Gardens, Kew.

Ebrahimzadeh, M. A., Hosseinimehr, S. J., & Gayekhloo, M. R. (2004). Measuring and comparison of vitamin C content in citrus fruits: introduction of native variety. *Chemistry: An Indian Journal*, *1*(9), 650-652.

Ehler, J. T. (2002). Key lime (Citrus aurantifolia). Food Reference, Key West, FL.

FAO. (2008). FAOSTAT. Satistical database

Fernandez-Lopez, J., Zhi, N., Aleson-Carbonell, L., Perez-Alvarez, J. A., & Kuri, V. (2005). Antioxidant and antibacterial activities of natural extracts: application in beef meatballs. *Meat Science*, *69*(3), 371-380.

Gulsen, O., & Roose, M. L. (2001). Lemons: diversity and relationships with selected Citrus genotypes as measured with nuclear genome markers. *Journal of the American Society for Horticultural Science*, 126(3), 309-317.

Hilu, K. W. L. A., & Liang, H. (1997). The matK gene: sequence variation and application in plant systematics. *American Journal of Botany*, *84*(6), 830-830.

Jayaprakasha, G. K., & Patil, B. S. (2007). In vitro evaluation of the antioxidant activities in fruit extracts from citron and blood orange. *Food Chemistry, 101*(1), 410-418.

Davis, Leonard. *Basic methods in molecular biology.* Elsevier, 2012.

Johnson, L. A., & Soltis, D. E. (1994). matK DNA sequences and phylogenetic reconstruction in Saxifragaceae s. str. *Systematic Botany,* 143-156.

Katzer, G. (2002). Lime [Citrus aurantifolia

Liogier, A. H., & Martorell, L. F. (2000). *Flora of Puerto Rico and adjacent islands: a systematic synopsis.* La Editorial, UPR.

Little Jr, E. L., & Frank, H. Wadsworth. 1964. Common trees of Puerto Rico and the Virgin Islands. *US Department of Agriculture, Agriculture Handbook,249,* 428-429.

Mabberley, D. J. (2008). *Mabberley's plant-book: a portable dictionary of plants, their classifications and uses* (No. Ed. 3). Cambridge University Press.

Manthey, J. A., & Grohmann, K. (1996). Concentrations of hesperidin and other orange peel flavonoids in citrus processing byproducts. *Journal of Agricultural and Food Chemistry, 44*(3), 811-814.

Morton, J. F. (1987). *Fruits of warm climates.* JF Morton.

Penjor, T., Yamamoto, M., Uehara, M., Ide, M., Matsumoto, N., Matsumoto, R., & Nagano, Y. (2013). Phylogenetic relationships of Citrus and its relatives based on matK gene sequences. *PloS one, 8*(4), e62574.

Swingle, W. T., Reece, P. C., Reuther, W., Webber, H. J., & Batchelor, L. D. (1967). The citrus industry. *The Citrus Industry, 1.*

Philippe, H., & Forterre, P. (1999). The rooting of the universal tree of life is not reliable. *Journal of Molecular Evolution, 49*(4), 509-523.

Hilu, K. W. L. A., & Liang, H. (1997). The matK gene: sequence variation and application in plant systematics. *American Journal of Botany, 84*(6), 830-830.

Bawden, A. L., Glassberg, K. J., Diggans, J., Shaw, R., Farmerie, W., & Moyer, R. W. (2000). Complete genomic sequence of the *Amsacta moorei* entomopoxvirus: analysis and comparison with other poxviruses. *Virology, 274*(1), 120-139.

McGeoch, D. J., Cook, S., Dolan, A., Jamieson, F. E., & Telford, E. A. (1995). Molecular phylogeny and evolutionary timescale for the family of mammalian herpesviruses. *Journal of Molecular Biology, 247*(3), 443-458.

Davison, A. J. (2002). Evolution of the herpesviruses. *Veterinary Microbiology, 86*(1), 69-88.

Sugita, M., Shinozaki, K., & Sugiura, M. (1985). Tobacco chloroplast tRNALys (UUU) gene contains a 2.5-kilobase-pair intron: an open reading frame and a conserved boundary sequence in the intron. *Proceedings of the National Academy of Sciences, 82*(11), 3557-3561.

Chen, B. Y., Janes, H. W., & Chen, S. (2002). Computer programs for PCR primer design and analysis. *PCR Cloning Protocols*, 19-29.

Timmer, L. W., Solel, Z., Gottwald, T. R., Ibanez, A. M., & Zitko, S. E. (1998). Environmental factors affecting production, release, and field populations of conidia of *Alternaria alternata*, the cause of brown spot of citrus. *Phytopathology, 88*(11), 1218-1223.

Cunningham, G. P., & Harden, J. (1998). Reducing spray volumes applied to mature citrus trees. *Crop Protection, 17*(4), 289-292.

Ghasemi, K., Ghasemi, Y., & Ebrahimzadeh, M. A. (2009). Antioxidant activity, phenol and flavonoid contents of 13 citrus species peels and tissues. *Pakistan Journal of Pharmaceutical Science, 22*(3), 277-281.

Manthey, J. A., & Grohmann, K. (1996). Concentrations of hesperidin and other orange peel flavonoids in citrus processing byproducts. *Journal of Agricultural and Food Chemistry, 44*(3), 811-814.

Manthey, J. A., & Grohmann, K. (2001). Phenols in citrus peel byproducts. Concentrations of hydroxycinnamates and polymethoxylated flavones in citrus peel molasses. *Journal of Agricultural and Food Chemistry, 49*(7), 3268-3273.

Anagnostopoulou, M. A., Kefalas, P., Papageorgiou, V. P., Assimopoulou, A. N., & Boskou, D. (2006). Radical scavenging activity of various extracts and fractions of sweet orange peel (Citrus sinensis). *Food Chemistry, 94*(1), 19-25.

Orwa, C., Mutua, A., Kindt, R., Jamnadass, R., & Simons, A. (2009). Agroforestree database: a tree species reference and selection guide version 4.0. *World Agroforestry Centre ICRAF, Nairobi, KE.*

Penniston, K. L., Nakada, S. Y., Holmes, R. P., & Assimos, D. G. (2008). Quantitative assessment of citric acid in lemon juice, lime juice, and commercially-available fruit juice products. *Journal of Endourology, 22*(3), 567-570.

Penjor, T., Yamamoto, M., Uehara, M., Ide, M., Matsumoto, N., Matsumoto, R., & Nagano, Y. (2013). Phylogenetic relationships of Citrus and its relatives based on matK gene sequences. *PloS one, 8*(4), e62574.

Johnson, L. A., & Soltis, D. E. (1995). Phylogenetic inference in Saxifragaceae sensu stricto and Gilia (Polemoniaceae) using matK sequences. *Annals of the Missouri Botanical Garden*, 149-175.

Steele, K. P., & Vilgalys, R. (1994). Phylogenetic analyses of Polemoniaceae using nucleotide sequences of the plastid gene matK. *Systematic Botany*, 126-142.

Liang, H., & Hilu, K. W. (1996). Application of the mat K gene sequences to grass systematics. *Canadian Journal of Botany, 74*(1), 125-134.

Gadek, P. A., Wilson, P. G., & Quinn, C. J. (1996). Phylogenetic reconstruction in Myrtaceae using matK, with particular reference to the position of Psiloxylon and Heteropyxis. *Australian Systematic Botany, 9*(3), 283-290.

Krishnakumar, K. N., Rao, G. P., & Gopakumar, C. S. (2009). Rainfall trends in twentieth century over Kerala, India. *Atmospheric Environment, 43*(11), 1940-1944.

Shyamalamma, S., Chandra, S.B.C., Hegde, M., & Naryanswamy, P. (2008). Evaluation of genetic diversity in jackfruit (*Artocarpus heterophyllus* Lam.) based on amplified fragment length polymorphism markers. *Genetics and Molecular Research*, 7(3), 645-656.

Xiaoming, C. Y. Y. H. P., & Xiuxin, G. W. D. (2001). An efficient method for genomic DNA extraction from woody fruit plants. *Journal of Huazhong Agricultural*, 5, 019.

Thompson, C. T., & Dvorak, J. A. (1989). Quantitation of total DNA per cell in an exponentially growing population using the diphenylamine reaction and flow cytometry. *Analytical Biochemistry, 177*(2), 353-357.

Muller, D., Lievremont, D., Simeonova, D. D., Hubert, J. C., & Lett, M. C. (2003). Arsenite oxidase aox genes from a metal-resistant β-proteobacterium. *Journal of Bacteriology*, 185(1), 135-141.

Bafeel, S. O., Arif, I. A., Bakir, M. A., Khan, H. A., Al Farhan, A. H., Al Homaidan, A. A., ... & Thomas, J. (2011). Comparative evaluation of PCR success with universal primers of maturase K (matK) and ribulose-1, 5-bisphosphate carboxylase oxygenase large subunit (rbcL) for barcoding of some arid plants. *Plant Omics, 4*(4), 195.

YOUR KNOWLEDGE HAS VALUE

- We will publish your bachelor's and
  master's thesis, essays and papers

- Your own eBook and book -
  sold worldwide in all relevant shops

- Earn money with each sale

Upload your text at www.GRIN.com
and publish for free